New Media

Behind the Iron Curtain

WYDAWNICTWO
UNIWERSYTETU
ŁÓDZKIEGO

Piotr Sitarski, Maria B. Garda, Krzysztof Jajko
with a Foreword by Graeme Kirkpatrick

New Media
Behind the Iron Curtain

Cultural History of Video, Microcomputers
and Satellite Television in Communist Poland

WYDAWNICTWO
UNIWERSYTETU
ŁÓDZKIEGO

JAGIELLONIAN
UNIVERSITY
PRESS

Łódź–Kraków 2020

Piotr Sitarski, Krzysztof Jajko – University of Łódź, Faculty of Filology
Institute for Contemporary Culture, Department of Film and Audiovisual Media
171/173 Pomorska St., 90-236 Łódź

Maria B. Garda – University of Turku, Artium, Centre of Excellence in Game Culture Studies
Sirkkala campus, Kaivokatu 12, FI-20500 Turku, Finland

Published by Łódź University Press & Jagiellonian University Press

First edition, Łódź–Kraków 2020

ISBN 978-83-8220-199-4 – paperback Łódź University Press

ISBN 978-83-233-4871-9 – paperback Jagiellonian University Press

ISBN 978-83-8220-200-7 – electronic version Łódź University Press

ISBN 978-83-233-7131-1 – electronic version Jagiellonian University Press

This book is published as a part of research project funded by the Polish National Science Centre,
decision number: DEC-2012/07/B/HS2/00419

Łódź University Press
8 Lindleya St., 90-131 Łódź
www.wydawnictwo.uni.lodz.pl
e-mail: ksiegarnia@uni.lodz.pl
phone +48 42 665 58 63

Distribution outside Poland
Jagiellonian University Press
9/2 Michałowskiego St., 31-126 Kraków
phone +48 (12) 631 01 97, +48 (12) 663 23 81, fax +48 (12) 663 23 83
cell phone: +48 506 006 674, e-mail: sprzedaz@wuj.pl
Bank: PEKAO SA, IBAN PL 80 1240 4722 1111 0000 4856 3325
www.wuj.pl

The book is available in the Columbia University Press catalog: https://cup.columbia.edu

TABLE OF CONTENTS

Acknowledgements

We would like to thank our colleagues who at various stages participated in our research and inspired us, especially Graeme Kirkpatrick and Piotr Witek.

The book emerged as a result of many interviews, not all of them are credited in the book but we are indebted to all people who shared their memories and thoughts with us; also to our students who conducted some of the interviews; and our friends and colleagues who helped us finding new media history respondents.

We are grateful to Kamila Rymajdo who was a careful and committed editor and devoted much time and effort to improve this book.

Our research and this book were funded by the Polish National Science Centre (grant number 2012/07/B/HS2/00419).

Piotr Sitarski
Maria B. Garda
Krzysztof Jajko

Foreword

Graeme Kirkpatrick

Viewed from the United Kingdom, where I lived at the time, Poland in the 1980s was a peculiar place. It was in the news as the country with an avowedly socialist government being challenged by a heroic trade union called "Solidarity." It was on movie screens in *Moonlighting* (Jerzy Skolimowski, 1982), a film in which Jeremy Irons played a Pole working illegally in Britain and watching, on multiple TV sets in a shop window, scenes of chaos in the streets of his home town. Part of the poignancy of this scene was the sense of distance and remoteness – the shop is closed, the screens are on the far side of the glass which Irons wipes furiously in an attempt to see better – conjoined with one of intimacy and closeness – the streets are familiar, his family are "there" a few feet away, yet all flights home have been suspended due to the military coup.

In the 1980s Britain had an oppressive government of its own, engaged in the violent repression of more than one trade union. Poland seemed similar, familiar even, yet all the terms through which such conflicts are normally interpreted (in particular, left/right) seemed to be reversed, as if in a mirror.

All around the world at this time new technological media were springing into existence. In Britain, the TV shop also doubtless contained VCRs (indeed, in one scene we see a sign advertising "video recorders"), TV satellite dishes and possibly some of the new home computers then being manufactured in the UK (Spectrums, Dragons, Acorns, BBCs). The reception and social shaping of these technologies in the Western context has been well studied, especially by scholars with a constructivist outlook. The discipline of Science and Technology Studies (STS) has explored the social shaping of home computers,[1] for example, showing how they were caught up in and shaped by marketing strategies with various, competing constituencies in mind.

The situation in Poland, which is presented in this book, was very different and yet in many ways also similar. The first appearance of new media was in a context largely free of marketing. In constructivist terminology, the new technologies were "underdetermined" everywhere – Andrew Feenberg writes that home

---

[1] Tom Lean, *Electronic Dreams: How 1980s Britain Learned to Love the Computer* (London: Bloomsbury Sigma, 2016).

computers, for example, "were launched on the market with infinite promise and no applications."[2] But in Western contexts some attempt was made to shape perceptions and to guide first users into certain kinds of relationship with the machines. Most commonly, educational uses were foregrounded and people were encouraged to buy home computers to give their children competitive advantage in the high-tech jobs market of the future. A concerted effort was made to cultivate a market for games played on the machines. This amounted to a significant cultural intervention, in which magazine publications, often with the backing of computer manufacturers, sought to define new kinds of consumer, capitulating them into participation in "gaming" and defining them as "gamers."[3]

This kind of cultural activity had implications for the way that distinctions might be drawn between objects, with some machines associated with learning (the BBC Micro in the UK, the Meritum computer in Poland) and social boundaries formed between "in" and "out" groups – in the case of gamers, this involved excluding anyone who wasn't young and male.[4] At the same time regulative concepts[5] were introduced, so that a new lexicon of appreciation ("great gameplay," "super graphics") came into circulation. In this way new fields of cultural activity were designed and new subjectivities ("gamer," "user," "hacker") were produced. In the British context, these processes were played out in the pages of magazines, at marketing events up and down the country, in school classrooms, on TV screens and in radio broadcasts, as well as in living rooms and teenagers' bedrooms.

In Poland, much of this activity was missing and the vital mediations that frame a new media technology, giving its meaning and locating it in its specific cultural niche were all very different. In each of the three chapters that follow, these issues are explored in connection with video, computers and satellite TV. These technologies were all equally "new" in Poland at this time, as elsewhere. The book rightly presents itself, therefore, as a study of "new media" that includes digital devices and challenges the easy, deterministic association of that phrase with computers, the internet and other "digital" innovations. The social and cultural context makes what it will of the affordances of each device. From the perspective of an age in which we think we know what computers and satellite TV "are" and what they are "for," it is fascinating to look back and reflect on the diverse range of our initial responses and uses.

---

[2] Andrew Feenberg, *Questioning Technology* (London: Routledge, 1999), 85.

[3] Graeme Kirkpatrick, *The Formation of Gaming Culture* (Basingstoke: Palgrave-Macmillan, 2015).

[4] Graeme Kirkpatrick, "Meritums, Spectrums and Narrative Memories of 'pre-virtual' Computing in Cold War Europe," *Sociological Review* 55, no. 2 (2007).

[5] Lydia Goehr, *The Imaginary Museum of Musical Works: An Essay in the Philosophy of Music. Revised Edition* (New York: Oxford University Press, 2007).

This book studies home computers and digital storage devices as part of a wave of new media technologies that also included video and audio tapes and associated machines (players and recorders) and new kinds of TV. All of these devices appeared a few years later in Poland than in Western Europe and the US, video arriving only around 1980, for instance, when it had already transformed popular culture in other parts of Europe. But the new media included things that didn't succeed elsewhere, like communal satellite TV services, and inventions, like teletext, that specifically didn't work in Poland even though they succeeded elsewhere. These things developed, or failed to, alongside magnetic video tape and other things that didn't catch on anywhere. One of the strengths of this book, which makes it such a delightful read, is the feeling of chaotic technical experimentation with objects and practices whose meaning has not yet been settled, in a context where the authorities tried to be controlling and restrictive but were mostly incapable of understanding what they should do in order to achieve this.

The focus of each of the studies included here is not on a linear "diffusion" of technology and its cultural impacts, but rather on the creative practices and associated cultural mediations that arose specifically during the early period of new media technology's presence in the Polish context. This culture of experimentation produced quirky practices that may be distinctive to the Polish context. For example, Piotr Sitarski describes how VHS was used for sound recording, when facilities designed for that purpose were unavailable. Was there a politics to this? Sitarski writes that many of these practices "existed on the margin of what was allowed." The government planned to introduce 2,000 video clubs, for example, where people could make and show films, but the people had other ideas, acquiring VCRs "under the counter" and organising private screenings, especially among student groups. Dynamism and responsiveness were largely the preserve of such informal group responses, as opposed to the seeming sclerosis of the administration.

This is nowhere more clear than when we read about the government trying to shape the new technology in line with some conception of socialism. Sitarski reports, for example, that in the 1980s thousands of VCRs were produced in Poland, most of them not sold to private individuals, but to state-run companies, cultural centres and universities. The socialistic aspiration implicit in this foundered on systemic incompetence, reflected in the decision to keep producing VCRs long after anyone had ceased making compatible tapes. And then there are the stories of a secret distribution network servicing the Party leadership with video entertainments, compounding the impression of a system that was failing on its own terms.

The theme of multiple, sometimes conflicting forms of reception is continued in Maria B. Garda's contribution, where she notes that party and state incorporated micro-computers in one way, while ordinary people did it through their own, multifarious forms of practice with varying degrees of institutionali-sation. The official discourse in the socialist Eastern Bloc countries focused on communal and not individual access to technology. But in Poland, as elsewhere, individual hobbyists desired to own their own computers and they were often willing to go to extraordinary lengths to obtain one.[6] In Poland, Garda estimates that about 80% of the computers in people's homes came from abroad, as people would acquire them when on visits overseas, or through the black market. This became a cause for concern for the authorities, who believed the machines were being used by spies.

Similarly, in the case of satellite TV, initially the authorities reacted with fear and tried to close down popular usage, before they realised that its expense alone was sufficient to keep the numbers involved very low. Meanwhile, there were plans to bring in a regulated version of satellite TV use across the Soviet Union. These plans were never implemented but there was an official use of sat-ellite TV, which involved state broadcasters recording programmes received from satellites and then re-broadcasting them, on official, terrestrial channels. Krzysztof Jajko's chapter tells the story of Porion, a company that import-ed satellite TV antennae into Poland from Sweden in the second half of the 1980s. The "repressive socialist state" effectively turned a blind eye but one gets the impression this was not benign but rather a reflection of complete ineffi-ciency: the socialist security apparatus was so atrophied it was simply incapable of assessing new threats.

At the same time, this was a state that was capable of supporting initiatives and even encouraging entrepreneurialism, as reported by one of Jajko's respon-dents, who recalls government-sponsored "research and development units," which was a status conferred on some businesses that would entitle them to sup-port, including tax exemptions: "In spite of how it seems, the way that everyone believes that everything that was back in the day, during so-called 'komuna', is so bad, it is not that way at all. Because there were also wise people. This is seen in the example of these research and development units. There was a government bill passed for those." Partly on this basis, socialist Poland became an exporter of satellite dishes.

Similarly, the authorities seem to have encouraged local, cable TV including associated social activism, which reflected enthusiasm for the TV and, as Jajko puts it, "could have been used for building a mature civil society" (p. 95), but was stymied by a lack of public resources to pay for cable. The government allowed private, local TV firms to sell satellite TV to housing co-ops, with representatives

---

[6] See Jaroslav Švelch, "Say it with a Computer Game: Hobby Computer Culture and the Non-entertainment Uses of Homebrew Games in the 1980s Czechoslovakia," *Game Studies* 13, no. 2 (2013).

from the co-ops choosing what would be broadcast to the whole community. In so doing they hoped to prevent what they perceived as more threatening, namely, the "free for all" of individual viewers choosing what they wanted to watch.

This book is an outcome of a several years-long research project, funded by the Polish National Science Centre. It represents one of several such explorations of the local contexts in which digital communications technologies were shaped by individuals and communities with diverse interests, who found multiple meanings in the technology. Similar studies are now appearing or have appeared of other local histories, notably of Australia and Czechoslovakia, for example, as well as my own work on the British context.[7] These studies undermine technology's appearance as the incarnation of authoritative knowledge and show that, in reality, it is the outcome of people experimenting and creating things that are meaningful to them. However, these social shaping processes do not occur in a vacuum but under conditions defined by different kinds of structural inequality and political domination. In this, they do not only concern the past but have much to teach us about the present and how the future will be made.

Bibliography

Feenberg, Andrew. *Questioning Technology*. London: Routledge, 1999.

Goehr, Lydia. *The Imaginary Museum of Musical Works: An Essay in the Philosophy of Music. Revised Edition*. New York: Oxford University Press, 2007.

Kirkpatrick, Graeme. "Meritums, Spectrums and Narrative Memories of 'pre-virtual' Computing in Cold War Europe." *Sociological Review* 55, no. 2 (2007).

Kirkpatrick, Graeme. *The Formation of Gaming Culture*. Basingstoke: Palgrave-Macmillan, 2015.

Lean, Tom. *Electronic Dreams: How 1980s Britain Learned to Love the Computer*. London: Bloomsbury Sigma, 2016.

Švelch, Jaroslav. *Gaming the Iron Curtain*. Cambridge, MA: The MIT Press, 2018.

Švelch, Jaroslav. "Say it with a Computer Game: Hobby Computer Culture and the Non-entertainment Uses of Homebrew Games in the 1980s Czechoslovakia." *Game Studies* 13, no. 2 (2013).

Swalwell, Melanie. *Homebrew Gaming and the Beginnings of Vernacular Digitality*. Cambridge, MA: The MIT Press, forthcoming.

---

[7] Jaroslav Švelch, *Gaming the Iron Curtain* (Cambridge, MA: The MIT Press, 2018) and Melanie Swalwell, *Homebrew Gaming and the Beginnings of Vernacular Digitality* (Cambridge, MA: The MIT Press, forthcoming).

PART I

Piotr Sitarski

NEW MEDIA AND THE FALL
OF THE POLISH PEOPLE'S REPUBLIC

Introduction

The Faculty of Philology at University of Łódź was, until recently, dispersed over the city centre, located in small and large buildings, tenement houses, and even single floors or flats. Walking through the streets, you could come across students carrying dictionaries and lecturer notes discussing Old-Polish devotional literature. It made the life of academia fit nicely into the rhythm of everyday life, making Łódź – in a seemingly paradoxical way – similar to a medieval city.

Before moving to a new building, which was to house the entire faculty, the university administrators decided to inspect the existing equipment and dispose of old and redundant items. Since they were focused mainly on audio and video equipment, my colleagues and I decided to examine the apparatus kept in the dusty storerooms. As it turned out, a cabinet located in a room used by a technician from a neophilology institute housed a ZK 140T reel-to-reel tape recorder together with a collection of tapes. In the last two decades of the Polish People's Republic (PPR) such recorders were an important tool in language teaching, so finding one was not so surprising. Because the device was working, we played some of the tapes to satisfy our curiosity, and, to our surprise, we discovered that they were recordings of disco music from the 1980s.

This event will not be recorded in the history of the university, and there will be no trace of it in the archives. Maybe a conscientious historian will be able to determine, after many years, what kind of equipment was used by specific institutes and departments, but they will not be able to learn anything beyond their official, intended use, for learning languages and entertainment. However, this find synthesises numerous important questions presented within this book. How did it happen that university equipment was used for recording songs that most likely had no didactic application? How was the recording made? Did someone listen to it while working? Did they lend it to someone organising a party and forget to erase the tapes? Perhaps it was some sort of technical trial of the device? Furthermore, who recorded music instead of linguistic exercises? Were they technicians with easy access to the device, or professors who had the proper know-how? Or, were these songs, perhaps, recorded by students? These questions compose this book's main research goal: that is, to discover the real application of revolutionarily new media technology.

Media in the society

It is worth starting our work with a few general, arguably even obvious remarks on media *sensu largo* and how they functioned in the society. Media is used for communication. The term "communication" contains intentional ambiguity connected with two basic meanings of the word. Firstly, communication is a deeply human activity, based on sharing signs, which ultimately convey experience, emotions, and viewpoints. In a nutshell, communication is understood here as in the title of James W. Carey's famous book – simply as culture.[1] Media is, therefore, in the very heart of human culture. This can be understood in a general and metaphorical way, as well as in a specific and historical manner. On the one hand, it is how we use media to communicate that make us tangibly human. On the other, media places us in a cultural reality which is changeable and unique. Cited by Carey, John Dewey says: "Society not only continues to exist by transmission, *by* communication, but it may fairly be said to exist *in* transmission, *in* communication."[2] The quote pertains to the histories and everyday lives of numerous specific societies.

Communication is also transmission – and this is the second basic, "cybernetic" way this term may be understood. In this context, communication creates a network of exchanges of signs, information, and goods, which are moved from one place to another. Thus understood, communication can be achieved without building a community, without a homogenous mental map, without emotional engagement. Therefore, it can pertain not only to humans, but also to animals or machines.

These two meanings, although apparently contradictory, in fact complement each other. By turning on a tape or a DVD player, I begin to communicate with the device as we exchange information. The builders of the device also, in a sense, communicate with me. We collectively share a similar view on the device: together we assume the same general aim and method of using any given piece of technology, which helps the individual understand how to turn on and use the device. For example, I know what the arrows and other pictograms on the buttons mean. When an image appears on the screen its creators communicate with me and engage in complex ideological, aesthetic, and other types of relations.

---

[1] James W. Carey, *Communication as Culture. Essays on Media and Society* (New York: Routledge, 2009).

[2] John Dewey, *Democracy and education*, Project Gutenberg, accessed May 17, 2019, https://www.gutenberg.org/files/852/852-h/852-h.htm

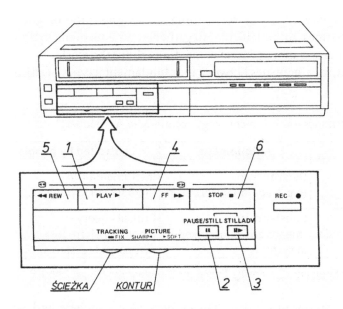

Odtwarzanie:
1 — start odtwarzania, *2* — stop-klatka, *3* — odtwarzanie poklatkowe (po uprzednim naciśnięciu klawisza PAUSE/STILL), *4* — przewijanie naprzód z podglądem, *5* — przewijanie wstecz z podglądem, *6* — stop

Fig. 1. Learning how to use buttons which seem intuitive and obvious nowadays

Source: Tadeusz Kurek, *ABC wideo* (Warszawa: Watra, n.d.), 58

Media, apart from being used in communication, are also machines. They belong to the same category as pickaxes, machine tools, or needles and threads. It is the reason why media is so often compared to simple tools, the reason why, in his famous book, Marshall McLuhan mentions clothes, accomodation and clocks next to newspapers and movies as examples of media.[3] Devices as tools exist in a twofold manner: they are a part of the material world, but they also belong to the social sphere. Lisa Gitelman refers to this when she defines media as "socially realised structures of communication, where structures include both technological forms and their associated protocols, and where communication is a cultural practice, a ritualised collocation of different people on the same mental map, sharing or engaged with popular ontologies of representation."[4]

The relationship between "technological forms" and "protocols," or, in other words, between technology and society, is one of the key questions both in media

---

[3] See Marshall McLuhan, *Understanding Media* (New York: Signet Books, 1964).
[4] Lisa Gitelman, *Always Already New* (Cambridge, MA: The MIT Press, 2006), 7.

theory, as well as in their everyday understanding. In fact, this question goes beyond theory and becomes an undeniable megatheory, an enormous metaphor joining views on media and more general convictions about the world.

Andrew Feenberg distinguishes between four fundamental types of attitudes towards technology:[5]

Tab. 1. Attitudes towards technology according to A. Feenberg

Technology is:	Autonomous	Controlled by people
Neutral (complete separation of means and ends)	Determinism (e.g. traditional Marxism)	Instrumentalism (liberal faith in progress)
Value-laden (means form a way of life that includes ends)	Substantivism (means and ends linked in systems)	Critical Theory (choice of alternative means-ends systems)

Source: Andrew Feenberg, *Questioning Technology* (London: Routledge, 1999), 9

If technology is autonomous, its changes depend on society, and while technological progress is driven by people, they can only perform tasks devised by technology. Feenberg sees traditional Marxism as an example of this attitude, believing that for Marx, technology is a neutral power, similar to the power of nature, one that acts continuously and can be used to different ends by a capitalist exploiter as well as by revolutionary proletariat: "The ultimate motive force of historical change is technology, productive forces, the whole of the equipment available to society plus acquired technical ability plus the technical division of labour."[6] Just like the proletariat revolution, technological development needs to take place; however, much like the revolution, it will not start itself. Through work in their laboratories and factories, scientists and engineers participate in a great project – the development of mankind. They are its agents, but they don't operate autonomously.[7]

Even though the table above presents the four types of attitudes as equal, determinism is, in fact, rarely accompanied by a belief that technology is neutral

---

[5] Andrew Feenberg, *Questioning Technology* (London: Routledge, 1999), 9.

[6] Leszek Kołakowski, *Main Currents of Marxism. Its Rise, Growth and Dissolution, Volume I: The Founders*, trans. P.S. Falla (Oxford: Clarendon Press, 1978), 337.

[7] This is Feenberg's position, but it is questionable to what extent Marx really perceived technology as a wholly neutral force. It is likely that the newest research on Marx's work, including unpublished works, will shed some light on the matter. See a detailed study on the role of technology in Marx's work: Regina Roth, "Marx on technical change in the critical edition," *The European Journal of the History of Economic Thought* 17, no. 5 (2010).

– that there is no connection between technological means and goals that can be achieved through them. In the case of Marxism, the development of technology – of the means of production – is a visible manifestation of historical progress and, as such, it is not neutral. Although for some time technology has been able to cause oppression, it must ultimately result in the triumph of the perfect society. This can be seen especially clearly in political practice, which *ex definitione* assumed the primacy of communist science and technology and saw them as a means of defeating their western competition. Technological progress has been seen not only as natural, and thus inevitable, but also desirable, since it hastened the victory of the "global proletariat." The belief in the inevitability and usefulness of technological progress was one of the secondary elements of practical Marxism-Leninism doctrine, but it was diffused through common awareness in Poland (and probably also in other countries of the Eastern Bloc) and became an important element of change.

Thus, Marxism sheds its belief in the neutrality of technology and changes into substantivism, which admits that technology is not just a flexible tool but also has its own substance which defines achieved goals. We can claim that firearms can be used in multiple ways (e.g. as starting pistols or flare-guns), but it can be reasonably argued that the substance of firearms boils down to shooting other people or animals, with other applications being marginal. Jacques Ellul, a radical substantivist, describes it like this:

> Technique integrates everything. It avoids shock and sensational events. Man is not adapted to a world of steel; technique adapts him to it. It changes the arrangement of this blind world so that man can be a part of it without colliding with its rough edges, without the anguish of being delivered up to the inhuman. Technique thus provides a model; it specifies attitudes that are valid once and for all. [...] But when technique enters into every area of life, including human, it ceases to be external to man and becomes his very substance. It is no longer face to face with man but is integrated with him, and it progressively absorbs him. In this respect, technique is radically different from the machine. This transformation, so obvious in modern society, is the result of the fact that technique has become autonomous.[8]

According to this view – shared by many modern thinkers who deal with media, especially the Toronto School headed by Marshall McLuhan – technology develops according to its own plan, and it cannot be used to serve just any goal. On the contrary, it sets its own goals, and new inventions and innovations trigger social changes: people adapt to the progress of technology. This approach

---

[8] Jacques Ellul, *Technological Society*, trans. John Wilkinson (New York: Knopf, 1964), 6.

usually has a gloomy and pessimistic dimension, despite originating from romantic resistance against industrialisation and its "dark infernal mills." In the twentieth century it was supported by the philosophy of Martin Heidegger, who in his essay The Question Concerning Technology describes the ways that technical thinking affects how modern people approach the world, how it becomes the main way to discover it, and how it changes the world into a technological resource. "A tract of land is challenged to putting out coal and ore. The earth now reveals itself as a coal mining district, the soil as a mineral deposit."[9]

Technopessimism, founded on the nineteenth-century fear of the peril that machines and their efficiency posed to humankind, together with a deterministic worldview, is not only important as an influential philosophical stance. It has also pervaded popular culture and become a foundation of, for instance, "science-fiction." Roger Caillois addresses this: "a science-fiction novel reflects the anxiety of our era, which is terrified at the very thought of the progress of technology theory, and which is no longer protected by science against the Unthinkable. Quite the contrary, it itself has started to pull mankind into the abyss. It is because science no longer means clarity and safety – it has become a disturbing mystery."[10] Even though this book is primarily meant to present views on the role of technology in our contemporary world, I mention the above because ideas have consequences. Ideas – embodied as they are in each situation – played an important role in the processes of innovation and its spread, described in this book. For this reason, one more thinker deserves our attention. He is much less well-known and influential than Martin Heidegger and mentioned much less often in the context of philosophy of technology, especially in English literature. Even though the belief in the substantial nature of technology is often pessimistic in form, with its vision of mankind enslaved by dark, antihuman forces, Pierre Teilhard de Chardin writes in his works about a very different world. As an evolutionist, de Chardin perceives the progress of technology as a consequence of a general process which leads from inorganic matter, to animals, to humans, to the creation of a collective awareness in the "noosphere." This process is not subject to human will; it is programmed into the world by God, and in this sense de Chardin is a substantivist. The evolution of mankind progresses to point Omega, where matter becomes purely spiritual, divine. This progress is, then, not only independent from humans but also leading to the ultimate good. According to de Chardin, the progress of technology – as a part of this larger process – is an element of a greater scheme which should not be opposed: "for Mankind as a whole,

---

[9] Martin Heidegger, The Question Concerning Technology And Other Essays, trans. William Lovitt (New York: Garland Publishing, 1977), 14.

[10] Roger Caillois, Obliques (Paris: Gallimard, 1987), 46.

a way of progress is offered and awaits us, analogous to that which the individual cannot reject without falling into sin and damnation."[11]

The ideas of Teilhard de Chardin became somewhat popular in the 1990s, because he was thought to have predicted the development of the Internet and inspired its makers.[12] Indeed, overwhelming evolution does entail, according to the French Jesuit, the creation of transistors and integrated circuits and, later – thanks to miniaturisation – personal computers. They, in turn, must connect into a network which will become the noosphere. These inventions, despite their human origin, had been predicted and programmed earlier in the divine plan and thus had to take place. The works of Teilhard de Chardin have been published, read, and commented on in Poland since the 1960s, and it is clear that he had some influence on shaping attitudes towards technology.[13] It may be no coincidence, rather a reflection of the zeitgeist, that the publication of the complete collection of his works began in Poland in 1984, at the time of new media revolution. The orthodox doctrine had a much more powerful impact, however, especially the teachings of the Catholic Church, which formed individual, as well as institutional, thinking. In the Feenberg table, it is situated in the neutral and non-deterministic area. The Catechism of the Catholic Church states it most clearly:

2293 Basic scientific research, as well as applied research, is a significant expression of man's dominion over creation. Science and technology are precious resources when placed at the service of man and promote his integral development for the benefit of all. By themselves however they cannot disclose the meaning of existence and of human progress. Science and technology are ordered to man, from whom they take their origin and development; hence they find in the person and in his moral values both evidence of their purpose and awareness of their limits.

2294 It is an illusion to claim moral neutrality in scientific research and its applications. On the other hand, guiding principles cannot be inferred from simple technical efficiency, or from the usefulness accruing to some at the expense of others or, even worse, from prevailing ideologies. Science and technology by their very nature require unconditional respect for fundamental moral criteria. They must be at the service of the human person, of his inalienable rights, of his true and integral good, in conformity with the plan and the will of God.[14]

---

[11] Pierre Teilhard de Chardin, *The Future of Man*, trans. Norman Denny (New York: Image Books/Doubleday, 2004), 10.

[12] See Jennifer Cobb Kreisberg, "A Globe, Clothing Itself with a Brain," *Wired* 1995, accessed December 12, 2016, https://www.wired.com/1995/06/teilhard/

[13] Teilhard De Chardin remains a disputed figure in Poland and in the last 50 years a number of books and papers discussing his thought have been published.

[14] Catholic Church, *Catechism of the Catholic Church*, 2nd ed. (Vatican: Libreria Editrice Vaticana, 2012).

It clearly follows that, first, technology (just like science) is subordinate to humankind. Second, *the Catechism* clearly states that moral values are external to research and engineering and they cannot be deduced from either usefulness or effectiveness. Of course, the writing comes from the 90s and so it came after most of the events presented in this book. However, it is based on the documents of the Second Vatican Council and from earlier ideas of the Catholic Church.[15]

An even clearer statement can be found in the *Compendium of the Social Doctrine of the Church*:

> 457. *The results of science and technology are, in themselves, positive.* "Far from thinking that works produced by man's own talent and energy are in opposition to God's power, and that the rational creature exists as a kind of rival to the Creator, Christians are convinced that the triumphs of the human race are a sign of God's grace and the flowering of His own mysterious design." The Council Fathers also emphasise the fact that "the greater man's power becomes, the farther his individual and community responsibility extends," and that every human activity is to correspond, according to the design and will of God, to humanity's true good. In this regard, the Magisterium has repeatedly emphasised that the Catholic Church is in no way opposed to progress, rather she considers "science and technology are a wonderful product of a God-given human creativity, since they have provided us with wonderful possibilities, and we all gratefully benefit from them." For this reason, "as people who believe in God, who saw that nature which he had created was 'good', we rejoice in the technological and economic progress which people, using their intelligence, have managed to create."[16]

This perspective presents technology not as a passive force that a man can use for any purpose, but rather as something positive which demands to be used.

A different approach is presented in other theories, which assume that although society controls technology, adopted technical solutions determine social life to some extent. The most interesting of these theories is known as Social Construction of Technology (SCOT) approach. Wiebe E. Bijker and Trevor Pinch, two of its main proponents, describe technical artefacts in terms

---

[15] See Vatican Council, *Pastoral constitution on the Church in the modern world: Gaudium et spes; promulgated by His Holiness Pope Paul VI on December 7, 1965* (Boston: Pauline Books & Media, 1998).

[16] Pontifical Council for Justice and Peace, *Compendium of the Social Doctrine of the Church*, accessed January 6, 2017, https://www.vatican.va/roman_curia/pontifical_councils/justpeace/documents/rc_pc_justpeace_doc_20060526_compendio-dott-soc_en.html

of interpretative flexibility: "By this we mean not only that there is flexibility in how people think of or interpret artifacts but also that there is flexibility in how artefacts are *designed*. There is not just one possible way or one best way, of designing an artifact."[17] This flexibility is not just about the ways that different created projects respond to different needs of different groups within society and that they, by competing, lead to various discoveries and inventions. Eventually the artefact is stabilised when one of its interpretations, suggested by a given social group, finally wins. At this point, a common understanding of what this artefact "is" can be established. Earlier problems are not solved in the technical, but rather in the social sphere. Namely, appropriate social groups should believe that difficulties have been overcome. This can be achieved by means of rhetoric, as in the change of the name (in the chapter devoted to VHS players I will describe how the "VHS tape" went through a phase of "cassette television" and finally became "home theatre"). Another way of ending the discussion is to redefine the problem. In this way, DVD finally defeated VHS: all problems associated with VHS players recording television signals were ignored and DVD players were presented as devices providing their users with perfect quality image and sound.

SCOT possesses useful tools for describing the progress of technology, which can be used even without accepting their theoretical background. The interpretation of an artefact and "closing" the debate in the phase of a "black box" does not have to assume that technology is determined by society. It is enough to surmise that interpretative flexibility is not total and the ending point of the debate is not random. The victory of the VHS player as a home theatre system in the 80s can be seen as a result of the impact of social forces. However, it can also be seen as the inescapable conclusion of the influence of the new technology, which was impossible to include in the status quo of contemporary mass media. This is important as the proliferation of new media coincided with the fall of the communist regime. It is very difficult to ascertain, empirically, if these two processes were connected in any way, and almost impossible to say which one could be the result of which. It is tempting to claim that new media overthrew communism, but the view that this falling system was no longer capable of controlling the media flow of information is also plausible. The facts described in this book can always be interpreted in one of these two ways.

---

[17] Trevor J. Pinch and Wiebe E. Bijker, "The Social Construction of Facts and Artifacts: Or How the Sociology of Science and the Sociology of Technology Might Benefit Each Other," in *The Social Construction of Technological Systems*, eds. Wiebe E. Bijker, Thomas P. Hughes and Trevor Pinch (Cambridge, MA: The MIT Press, 2012), 34.

New media

Media constantly changes. No matter which view we adopt on the relationship between technology and social forces, the rapid flow of media is a fact. When analysing media reality, it is worth remembering that we are facing a process, and that if it is fragmented, and divided by moments of clearly defined, unchanging media, these moments are fleeting (if not illusory). In the flux of change we can distinguish continuous and discrete elements. The path leading from a cable telegraph to wireless telegraphy, to radio, to television, and to satellite television can be seen as one flux without any discrete changes, and we can distinguish between these media only on the basis of some incoherent conventions. For example, the telegraph differs from early radio communication mostly in how it was used, rather than in terms of technology employed. The wireless telegraph was used for "serious" matters by businesses, governments, the military, and early radio communication by amateurs, usually as a hobby. When it comes to the television and the radio, however, they are (and have been) used in almost identical ways: as forms of entertainment. The flux of media can also be divided by paying attention to discrete elements which suspend the flexibility of interpretation for some time and place the media in the social sphere as an apparently unchanging and obvious artefact. The permanence of naming is an important element of the "black box" phase introduced in the SCOT. For instance, we can claim that a radio that stands on our desk is different from its mobile version we carry in our pocket, but the institutional stability of radio broadcasters led to people seeing the technical change as an improvement, not a break in continuity.[18]

How can we define new media with all these obstructions and technicalities at hand? The "narrow devotion to the present," as Lisa Gitelman and Geoffrey B. Pingree put it, prevails in contemporary academic and, especially, common discourses.[19] Its cornerstone is absolutizing novelty, most often accompanied by identifying it in an essentialist manner with the characteristics of computer media. The belief that new media are digital media, as suggested by Lev Manovich in his book *The language of new media*, is an obvious example of this. He avoids using the phrase itself, replacing it with the rule of "numeric representation," but this is his way of making the term more precise. Such a definition excludes magnetic

---

[18] It is partly connected with technological change, namely, with using transistors instead of vacuum tubes, even though there were tube-based radio receivers (e.g. the model Szarotka produced in Poland in the 1950s or Tesla Minor from Czechoslovakia).

[19] Lisa Gitelman, Geoffrey B. Pingree, *New Media, 1740–1915* (Cambridge, MA: The MIT Press, 2003), xi.

tape players, which the scholar classifies as "old" devices, and mixed technologies
– analogue-digital, like the Laser Disc.[20]

According to the logic of novelty, such a view seems outdated and too broad
today. Progress is unstoppable and for twenty-first century scholars, novelty
in media is usually no longer about whether they are digital, but rather about
their social potential. For instance, Sean Cubitt in his erudite article "Media
Studies and New Media Studies," defines the moment of birth of new media as
13 October 1993, the day the internet browser Mosaic premiered. He admits that:
"Other dates might work as cleanly – the personal computer revolution of the
1980s, perhaps – but do not entail the common-sense awareness that emerged
in the ensuing months that something massive and life-changing had begun."[21] As
a result, Cubitt considers everything that happened earlier to be prehistory: "Pri-
or phases, as far as mainstream media studies is concerned, constitute a prehistory of the popular or mass uptake (depending on the school of thought involved)
that turned laboratory or experimental formats into technical media of the scale
and significance of the press or television. 1993 can, then, serve as the watershed
of the new in media studies."[22] More pessimistically inclined authors such as Andrew Keen have criticised the transformation of media, seeing the essence of this
transformation as communal.[23]

Such an outlook on new media evidently narrows their definition: now
it is not enough to have numeric representation, and only a part of the Internet
is considered a new medium. Simultaneously, the emphasis also shifts and presents a slightly different line of progress for these media, one more connected with
the processes of networking and interaction between their users than with the numeric representation of new media objects.[24] It is worth noting how this current
definition changes the perception of historical development. Popular history

---

[20] See Lev Manovich, *The Language of New Media* (Cambridge, MA: The MIT
Press, 2001), 46.

[21] Sean Cubitt, "Media Studies and New Media Studies," in *A Companion to New
Media Dynamics*, eds. John Hartley, Jean Burgess, and Axel Bruns (Malden: Wiley-Blackwell, 2013), 16.

[22] Ibid. The research presented in this book should, therefore, be classified as pre-
historic.

[23] See Andrew Keen, *The Cult of the Amateur: How Today's Internet is Killing Our
Culture* (New York: Doubleday/Currency, 2007).

[24] The idea of seeing new media as networked is developed mainly by Manuel
Castells. See his *The Internet Galaxy* (Oxford: Oxford University Press, 2001), and *The
Rise of the Network Society*, 2nd ed. (Malden: Wiley-Blackwell, 2010). Henry Jenkins,
on the other hand, perceives convergence to be their characteristic feature. See Henry
Jenkins, *Convergence Culture: Where Old and New Media Collide* (New York: New York
University Press, 2006).

of social media seldom mentions Charles Babbage or the Turing machine, but often includes obscure Bulletin Board Systems (BBS).

In each of these new media descriptions, the content of the definition changes; but what remains constant is the modernist structure of excitement over the very essence of novelty. For that reason, the term "new media" has remained attractive over the years. First, it is the direct result of the ideology of progress. The change is not about simple replacement, it has an axiological dimension – it is advantageous and inevitable. Such an attitude, coupled with a search for the essence of novelty, results in the replacement of historical thinking with historiosophical thought. The latter treats the progress of media as a one-way process driven by some general rule, whose final realisation – which may not be surprising – turns out to be close at hand. This is exemplified by Howard Rheingold's idea of virtual reality as the ultimate goal of media progress:

> We are approaching a breakpoint where the quantitative improvement in that model-building interface will trigger a qualitative quantum leap. In coming years, we will be able to put on a headset, or walk into a media room, and surround ourselves in a responsive simulation of startling verisimilitude. Our most basic definitions of reality will be redefined in that act of perception.[25]

We can see one more characteristic of historiosophical thinking here: the necessity of revolution, of breaking continuity, of clearly distinguishing between the old and the new. This corresponds with consumerist culture, driven by the need for constant change of goods, which justifies circulation: purchasing a product is crucial because the previous product, although still efficient, is no longer new. As one old business proverb, often attributed to Marshall McLuhan, goes: if it works, it's obsolete.

Despite these critical remarks, I do not think that the rejection of the term "new media" was right; but at the same time, I am not absolutizing this term. It has indeed structured the thinking of scholars and ordinary users, which makes it interesting as research material, and although its repetitive application to endlessly new phenomena can be tiring or even funny, its constant form contains a new reality with each such application. Each new "new medium" is different from the previous ones, and if we return to its historical meaning it becomes a useful tool for describing a given media situation. It should be investigated, then, what meaning the term conveys at any given time, how it is understood, and how it organises the media experience.

In a way, this whole book is devoted to this task. In order to specify the scope of this research however, we should begin with how new media were perceived

---

[25] Howard Rheingold, *Virtual Reality* (London: Mandarin, 1991), 388.

in the last years of the Polish People's Republic. This will allow us to describe the media landscape in its own right, without placing it in some broader historiosophical context and losing sight of the dynamics of change. In other words, it is not about what we can consider new from our perspective, but rather that which was perceived as new at the time. Some of those new media no longer exist or vegetate without much importance and/or publicity; some others have been transformed. All of them, however, deserve attention, not because of their place in some hypothetical chain of evolution, but rather because of the interest they sparked at the time. Media archaeology teaches us that coming back in time to the moment when things were novel allows us to see unexpected connections. Despite the apparent triumph of VHS tapes over laser discs, video tape players are no longer manufactured, and DVDs continue to fare well. Videotext seemed very promising at the end of the 1970s, but a few years later it was hailed as a failure when faced with computer messaging. In the end, videotext survived and became a ubiquitous and indispensable element of radio and TV broadcasts.[26]

Even though numerous new media have appeared in the cultural landscape since the nineteenth century, it was only in the second half of the twentieth century when the term was introduced and widely accepted in both common and academic discourses. It was connected both with the rise of consumerism, and the development of research on communication and other fields, such as sociology or management studies, which included media in their scope of interest. The American information service DIALOG, which started in 1966 and included over 200 bibliographic databases in the 1970s, saw a gigantic increase in the number of publications devoted to media.[27] In the year 1974, 39 articles on new media appeared in popular magazines, and in 1982 the number grew to 1,326 articles. At the same time, the number of articles in scientific magazines devoted to economy and management grew from two to 503.[28] Whatever theoretical doubts we might have regarding the terminology, new media became an important topic of public and scientific debate in the late 70s and early 80s (at least in the USA).

At that time, new media were most often defined in two ways: through their interactivity being the main characteristic of use and, on the technical

---

[26] On the failure of videotext see A. Michael Noll, "Videotex: Anatomy of a Failure," *Information & Management* 9, no. 2 (September 1985).

[27] It still operates, taken over by a different company, at http://www.proquest.com/products-services/ProQuest-Dialog.html, accessed January 15, 2017.

[28] The information is taken from Ronald E. Rice's article "Development of New Media Research," in Ronald E. Rice & Associates, *The New Media. Communication, Research and Technology* (Beverly Hills: Sage Publications, 1984), 16–17. The author also discusses certain limitations of this data.

side, through the use of computers. In his influential work from that period, Ronald E. Rice describes new media as "communication technologies, usually using the possibilities of computers (microchip and mainframe), which enable or facilitate interactivity between users or between users and information."[29] The idea of interactivity was later frequently analysed and criticised. Rice saw the term's lack of precision and drew our attention to the social dimension of interactivity, juxtaposing it with mass application. The way he understood it, new media, contrary to old media, is no longer mass media: they do not force their users into simultaneous reception of the same content.[30]

Although computers appear in this definition somewhat by accident, in the following technical description they are described as a "foundational element of new media."[31] Apart from that, Rice mentioned new media channels of transmission (coaxial cable, twisted-pair cable, optical fibre, microwave transmission, mobile phones, and telecom satellites) and the method of storing data (interestingly, he wrote about videodiscs but not magnetic tapes).

The authors of this work treated the "new media" label loosely, and their definition was used only to specify the scope of analysis. They understood new media in a relative way, and believed that their novelty was temporary: "these media are new only to the generation first experiencing them."[32] Therefore, the authors of the articles cited above narrowed or broadened the meaning of the term to suit their needs.

A similarly wide definition of new media can be found in another important work from that time: Everett M. Rogers' book *Communication Technology. The New Media in Society*. In it, he writes:

> The key technology underlying all other new communication technologies is electronics. Electronics technology these days allows us to build virtually any kind of communication device that one might wish, at a price [...]. One special characteristic of the 1980s is the increased number and variety of new communication technologies that are becoming available. Further, and more important, is the

---

[29] Ronald E. Rice, "New Media Technology: Growth and Integration," in Ronald E. Rice & Associates, *The New Media. Communication, Research and Technology*, 35.

[30] See ibid. Manovich, among others, conducts a critical analysis of interactivity in *The Language of New Media*. See also Yuping Liu, L. J. Shrum, "What is Interactivity and is it Always Such a Good Thing? Implications of Definition, Person, and Situation for the Influence of Interactivity on Advertising Effectiveness," *Journal of Advertising* 31, no. 4 (2002).

[31] Rice, "New Media Technology," 36.

[32] Ronald E. Rice, Frederick Williams, *Theories Old and New: The Study of New Media*, in Ronald E. Rice & Associates, *The New Media. Communication, Research and Technology*, 55.

nature of how these new media function; most are form many-to-many information exchanges. Their interactive nature is made possible by a computer element that is contained in these new technologies.[33]

Rogers began his definition with electronics, but ultimately focused on their interactive nature enabled by the use of computers.

In Poland, the term "new media" first appeared a few years later; in such a dynamic context this represents a significant delay, and shows our position in the hierarchy of countries undergoing technological changes. Obviously, the term "new media" was almost absent from the Polish popular press and cannot be compared to the popularity of this phrase in the West. Impacted by a tremendous economic crisis, communism in the last years of its existence within Poland was not willing to engage in marketing campaigns advertising the consumption of electronic goods, so no catchy slogans were necessary. Occasional and feeble actions taken in the context of VHS players and later PCs, did not result in the awareness of new media as a homogenous area of consumption (based on interactivity). It is a fact, however, that some new media functioned in the sphere of consumption: as a tool for entertainment, a source of prestige, or even as objects of accumulation. Although VHS players and computers were often mentioned in the popular and professional press, these phenomena were not presented under any one umbrella term. Some of the more technical magazines contained articles on computers and audio-visual technologies, but they perceived them simply as electronic devices rather than as new media. One example is the magazine *Audio-Video*, published since 1984, initially as a supplement of *Radioelektronik*, which carried the meaningful subtitle "Postępy w elektronice powszechnego użytku" ("Advances in Everyday Electronics"). Despite its name, *Audio-Video* contained articles on computers, which the editorial from the first issue justified as being relevant.[34] This promise was kept, and the magazine published articles devoted not only to cassette players and TV sets but also to VHS players, laser players, satellite television, and, most importantly, microcomputers. They were not, however, described as new media.

Audio-Video began publication in the mid-80s, but new media appeared in a narrower academic context much earlier. Articles on new media appeared mostly in *Przekazy i Opinie* magazine, published by OBOP (Polish Centre for Public Opinion Research) under the auspices of Radiokomitet.[35] The magazine

---

[33] Everett M. Rogers, *Communication Technology. The New Media in Society* (New York: The Free Press, 1986), 2–3.

[34] Jerzy Auerbach, "Od redakcji," *Audio-Video*, no. 1 (1984): 1.

[35] Radiokomitet (The Committee of Radio and Television) was the broadcasting regulatory authority set up by the communist government in 1951, with the power to

prioritised new media from the very start. The first issue contained Jerzy Kossak's "Innowacje elektroniczne i hipotezy zmian kulturalnych" ("Electronic Innovation and Cultural Change Hypotheses"), where the author discussed cultural changes that he predicted would take place because of technological advancements: video cassettes, satellite and cable television, the growth of electronic screens, mass production of electronic video cameras, popularity and miniaturisation of chips, and the use of holography.[36] Apart from translations and original articles, *Przekazy i Opinie* also contained a large number of reports and reviews, and was supposed to familiarise the readers with international technological novelties and writings (Poland's position as a peripheral country is evident here). The second issue of the magazine featured many articles on cable television and Television Electronic Discs.[37] In subsequent issues, new media topics were discussed systematically, and the term "new media" appeared explicitly as late as 1980. In the report section of issue 22 of the magazine there is an article titled "Dyskusja nad sprawą nowych mediów" ("A Discussion on the Matter of New Media"). It is a piece by Jerzy Roszkowski, composed of summaries of two articles published originally in the German magazine *Publizistik*.[38] Although the term "new media" appears in the title of the text, borrowed from one of the German articles, it does not seem to be particularly interesting to the author. Issue 26, however, includes a table of contents of all previous publications, and in the "New Media" category 16 articles are listed, most of them devoted to cable and satellite television.[39] In the following years, they appeared even more often: the table of contents from issues 26-52 notes 26 such pieces.[40] This is, overall, still a low number, especially if we consider that most of these articles are translations, reviews, and reports.

The picture inferred from published foreign news, reports, and original articles is rather inconsistent. News and reports usually pertain to cable, satellite television, and new aesthetic forms in television in general, such as video clips – which is understandable if we consider the general profile of *Przekazy i Opinie* and the public opinion research institution that published it. The authors of original papers did not suggest that new media can have any inherent common features either.

---

formulate broadcasting policy, to plan the strategic development of radio and (later) television, and to oversee the state broadcasters.

[36] See Jerzy Kossak, "Innowacje elektroniczne i hipotezy zmian kulturalnych," *Przekazy i Opinie*, no. 1 (1975).

[37] See Halina Żebrowska, "Nowości techniczne," *Przekazy i Opinie*, no. 2 (1975).

[38] Jerzy Roszkowski, "Dyskusja nad sprawą 'nowych mediów'," *Przekazy i Opinie*, no. 22 (1980).

[39] See *Przekazy i Opinie*, no. 26 (1981): 209.

[40] See *Przekazy i Opinie*, no. 53–54 (1988): 289–290.

Michał Gajlewicz's opinion is the closest we can find in terms of new media uniformity, as he is the author who dealt with the topic most frequently and tried to present an original definition of new media. Gajlewicz pointed to their interactivity as a shared characteristic, as opposed to passive and active perception. He also claimed that new media, "beside their specific functions, serve all those of the old media."[41] He also noted that "novelty" is more than a specific position on a timeline, giving the example of the telephone, which is an old medium but also an interactive one, and which is often used as a technological base for newer systems.[42] Gajlewicz also presented seven technical factors which affected the development of these media (the use of cable in data transmission, digitalisation of transmission, the use of satellites, the use of computers, tape video recording, the application of knowledge about human perception in the building process, lower costs of production, market penetration and economic rationalisation).[43] As we can see, the author thought that the use of computers was something different to the digitalisation of transmissions, and both factors were of the same importance as the use of tape in video recording (and not just for audio, as was the case in the past). Gajlewicz also enumerated new media, clarifying that his list was neither precise nor disjointed; he listed satellite television, cable television, pay per view videographic systems, video equipment, new editing techniques (including those using computers), and solutions for improving the quality of images and sounds, as well as those used for achieving the "effect of presence."[44]

Three years later, in an article titled "Nowe media – stare i nowe dylematy" ("New media – Old and New Dilemmas"), Tomasz Goban-Klas claimed that "We define new media by enumeration as all those means of communication (in the broadest sense) which make use of electronics, especially integrated circuits and digital signal encoding, for recording and broadcasting information."[45] He went on to list examples of new media, such as video, cable TV, and satellite transmission, and he characterised them as, among other things, interactive. In the same issue of *Przekazy and Opinie* there is also a text by Karol Jakubowicz using a relative definition, describing as new "all technologies of receiving, recording, processing, and broadcasting information, data, sounds and images that

---

[41] Michał Gajlewicz, "Nowe media," *Przekazy i Opinie*, no. 41–42 (1985): 282.

[42] See ibid. That remark is still valid since the majority of new services (for example Internet providing services) are based on traditional telephone lines.

[43] See ibid., 283–284.

[44] See ibid., 284–289.

[45] Tomasz Goban-Klas, "Nowe media – stare i nowe dylematy," *Przekazy i Opinie*, no. 51–52 (1988): 15.

were introduced after traditional television."[46] He also characterised new media, but did not mention their digital aspect; he saw computers as a means of enabling immediate access to information.

This account of new media definitions in the last years of the PPR is crucial if we are to realise how much commotion this phenomenon caused and how uncertain its fate was. When choosing media for detailed analysis, the authors focused on those that were perceived as "new" at the time. Their choice of personal computers and computer games; consumer video recording technologies; and cable and satellite television, is not shaped by some homogenous category they all belonged to. Admittedly, all of these media can be defined by their interactivity, but if we apply the term so broadly that it loses its power, we are left with the notion that these new media gave their users relatively more choice than their older counterparts. The point about relativity is important as far as the broadly understood interactivity of new media can be questioned by juxtaposing the gramophone (an old medium) with satellite television (a new medium), or by presenting the case of the telephone (as in Gajlewicz's article). Satellite television should then be compared with traditional (landline) television, and the VHS player with cinema.

An attempt at perceiving these media in a contemporary perspective requires abandoning the teleological approach. The choice we have made in this book is to examine the media that was successful from our perspective (such as PCs or computer games) as well as those which finally failed (magnetic tape video recording). We have not included, however, videotext and teletext, which were commonly considered to be new media in the 80s. This decision does not stem from the fact that they have been buried in the scrapheap of dead technologies, which is not necessarily true. Simply, neither teletext nor videotext were widely used in the PPR, and it would be difficult to describe them from the perspective of ordinary users. Although an experimental system was presented as early as 1980, television Telegazeta (a form of teletext) was first broadcast in 1988.[47] But this remark is a formality, especially if we realise that the timeframe used in this book is not very strict. The main problem we face when considering these media pertains to the analytical perspective we have adopted. Teletext, first broadcast by Polish Television and later by other networks, was used like other new media: it was broadcast as a mass medium but allowed its users to choose their own content. It could not be used in any extra-systemic way, though, and thus is largely absent from this book.

---

[46] Karol Jakubowicz, "Nowe techniki informacyjno-komunikacyjne: czy mogą stworzyć nowy układ kultury," *Przekazy i Opinie*, no. 51–52 (1988): 28.

[47] See Wikipedia, "Teletekst," accessed January 20, 2017, https://pl.wikipedia.org/wiki/Teletekst, and Krystyna Prószyńska, Marek Błachut, "Teletekst," *Radioelektronik* 123 (1989). The authors discuss the issue of displaying Polish diacritics by the teletext.

Extra-systemic media practices

Extra-systemic use of media, as one of the key conceptual elements of this book, requires explanation. The previously mentioned definition of media by Lisa Gitelman describes them as communicational structures: forms of technology and social protocols. It is a static definition, describing a medium as something that already exists and is relatively unchanging. Thus, it describes media normatively, explaining what they should be and how they should be used. Technology, as well as the protocols of its use, must be accepted, going through the phase of interpretational negotiations during which users establish how a given medium should be used and what it will become to them. The SCOT theory describes it as a top-down process and emphasises the influence of society. The notion of extra-systemic use does not reject this approach and refers to what specific users do with a given medium when using it to suit their own needs, consciously or unconsciously ignoring their protocols of use or just taking advantage of the fact that such protocols have not yet been established. Users are not aware of participating in any form of negotiations of meaning; they just play with the medium exploring its possibilities. Some of these uses can be more institutionalised and common, while others remain individual, but none of them falls within the existing, commonly accepted system of media use. Extra-systemic practices concentrate – from the perspective of their users – on the media, and any remaining aspects, especially political, are of secondary importance.

The concept of extra-systemic practices is fuzzy, as are the ideas of protocols and technological forms noted by Gitelman. Let us consider the following example. At the beginning of the 90s, Krzysztof Stasiak, a music producer, ran a recording studio in Upper Silesia. For his work, he used a Fostex tape recorder, copying the recorded audio material to a Hi-Fi stereo video cassette recorder.[48] This allowed him to achieve a high-quality recording without resorting to DAT technology, which was very expensive at the time. By doing so, he breached two protocols. The first pertained to the distinction between professional and amateur technologies, and commonly held convictions about how hugely different they are. It turns out, however, that the difference is sometimes more about the social protocol (the important elements of which are price and availability) rather than about the technological form (the quality of sound was comparable). The second breach was about using a video device for audio recording, which ignores the "basic" application of the device. The result of such a practice was that the prices of digital recorders fell to a level that allowed their everyday use,

---

[48] Krzysztof Stasiak, interview, February 26, 2016. The interviewee admits that he was inspired by an interview with Frank Zappa who mentioned a similar solution.

without resorting to VHS recording. Using VHS technology for audio recording did not become a media norm, remaining extra-systemic, but it still attracts some interest.[49] Such is the fate of most extra-systemic practices, which are, first and foremost, local (or even individual), and, second, usually short-term. The story of the professional use of amateur equipment is slightly different. This breach of protocol is a part of a broader trend that was not rejected. Quite the contrary, it was adopted and became similarly popular to practices such as digiscoping or using digital cameras for filming professional videos. Aside from these two possibilities, there is also the chance that an extra-systemic practice will become conventional, meaning that it will be accepted as a social or technological norm. A rather eccentric example of such an outcome was the foundation of Polska Partia Posiadaczy Magnetowidów (Polish Party of VCR Owners), officially registered in 1991. The goal of this pseudo-political group was to bypass the newly introduced regulations on copyright law, which made copying or renting VHS tapes more expensive. Unrestricted copying of tapes – extremely popular in the 80s – infringed upon the interests of distributors, who wanted the market to be more "normalised," which was, of course, in their own interest. The VHS rental owners responded by moving their businesses to the newly created political party – at least for a few years – because it legalised their business model (until 1994, when a special bill on copyright law was passed in the parliament).

The concept of extra-systemic practices is so important – especially now, under the influence of the Internet – because it has become clear that the assumed normative ways of using media are often ignored by their users. The Humanities realises this with greater clarity and hence we see more research on fan fiction and the work of fandom in general, such as the work of Henry Jenkins.[50] Ramon Lobato looks at the matter from a different, economic-legal perspective, and claims that in cinematography this practice is not marginal but rather a global standard against the regulatory expectations of big distributors.[51] This example takes my argument towards politics and economy but the concept of extra--systemic practices, as I see it, pertains mostly to what people do with the media; the political or economic aspects of their actions are of secondary importance.

---

[49] There is some information on the Internet regarding the use of Hi-Fi video recorders for sound processing. See Greatbear audio & video digitising, "Digitising Stereo Master Hi-Fi VHS Audio Recordings," accessed January 2, 2017, http://www.thegreatbear.net/audio-tape/digitising-vhs-hifi-audio-recordings. Some musicians still experiment with VHS compression in their recordings: Big's Gaming, "Audio Mastering – VHS Worth It?," accessed January 2, 2017, https://www.youtube.com/watch?v=7iNhzEpmz94

[50] See Jenkins, *Convergence Culture*.

[51] See Ramon Lobato, *Shadow Economies of Cinema. Mapping Informal Film Distribution* (London: BFI and Palgrave Macmillan, 2012).

This concept is thus distinguished from other perspectives, which see media practices as just an emanation of other more important and more powerful forces.

For example, the cultural reality of the PPR is often described in terms of an antithetic pair: power (the first, or official, circulation) versus opposition (the second circulation, or *samizdat*). This perspective has obvious advantages, encompassing most cultural activity in Poland in the communist era. There are, however, large areas beyond that dichotomy. This problem was brought to light in the monumental *Bibliografia publikacji podziemnych w Polsce* (*Bibliography of Underground Publications in Poland*), published in Paris, where Władysław Chojnacki and Wojciech Chojnacki (under the pseudonym Józefa Kamińska) wrote: "I have also omitted all pornographic publications, dream-books and typical tawdry literature, as well as books popularising herbal medicine, etc. published solely for commercial purposes."[52] The authors did not justify such an exclusion, but we can guess that it was because of the commercial (rather than ideological or political) character of these publications. Part III of the work contains a list of 99 audio cassettes. The authors admit that they had not been able to find all audio publications, but they did not stipulate whether they had been selective in presenting their findings (e.g. if there were any cassettes with dance music available in the second circulation at the time). Their approach to documenting video cassettes is even less detailed. The authors wrote: "I also have to say that three years ago video cassettes appeared in underground circulation. However, until this moment I have not been able to find these films apart from the first video cassette of Niezależna Oficyna Wydawnicza (Independent Publishing House) by Ryszard Bugajski, titled *Przesłuchanie* (*Interrogation*), recorded in 1985."[53] In the mid-80s Poland already had a well-developed market of VHS cassettes, although the authors did not consider it "underground," possibly due to its commercial nature. It is not my intention to diminish the value and importance of *Bibliografia*, which in other respects is thorough and reliable. But if we take its media part at face value, we should assume that in Poland in the mid-80s a hundred independent audio cassette titles were in circulation, together with an unspecific, but rather small number of films recorded on video cassettes. This is very distant from reality. Moreover, the second volume of *Bibliografia* ignores all other forms of publications except for books and magazines – this time without any explanation.[54] It does not mean that no such circulation existed in the years 1986–1987, quite

---

[52] Józefa Kamińska, *Bibliografia publikacji podziemnych w Polsce, 13 XII 1981 – VI 1986* (Paris: Editions Spotkania, 1988), 13.

[53] Ibid., 15.

[54] See Wojciech Chojnacki, Marek Jastrzębski, *Bibliografia publikacji podziemnych w Polsce. Tom drugi: 1 I 1986 – 31 XII 1987* (Warszawa: Editions Spotkania, n.d.).

the contrary – the video market was thriving to the extent that it exceeded the traditional formula for samizdat adopted by Chojecki and Jastrzębski.

The shortcomings of the official-oppositional dichotomy inspired the concept of "the third circulation." It refers mostly to the youth culture of the 80s, especially music and music subcultures.[55] Even though it is commonly used and describes an important area of the cultural landscape of the Polish People's Republic, its scope is relatively small: it does not include either pornography nor herbalism, not to mention science-fiction. Moreover, the third circulation is a term that is strictly connected to politics because it pertains to subcultures that were purposefully and consciously apolitical or even antipolitical (such as punk).

The term "extra-systemic practices" used in this book refers to the socio-political system of communist Poland, but it cannot be limited to this system, nor to any form of opposing it. The works of Krzysztof Stasiak described above were neither illegal nor political – their extra-systemic quality resulted from a novel use of a medium. Most actions described in this book were not political in character, nor were they consciously aimed against the communist government (even if they might have contributed to its demise). They existed on the margin of what was allowed and suggested, sometimes using the support of official institutions, such as clubs or community centres, taking place in private homes, or arising in spontaneously created environments.

The diffusion of new media in the Polish People's Republic

This book is about the last years of communist Poland. Since it describes media and not politics, the timescale here is not very precise. We begin our analyses with the first appearance of new media – if, of course, we have been able to find such moments and documents or people that would provide us with the relevant information. Much depended on the medium: personal use VHS players were first produced in Poland in 1973, while satellite television formally operated since 1975, although its first private application happened a decade later. Microcomputers arrived in Poland in the early 80s.

Establishing the ending stage is much more difficult. Even if we were to use political dating, finding the date of the Polish People's Republic final demise is not that obvious.[56] Media life, however, does not strictly follow political life, which is one of our main assumptions. Another date that comes to mind is February 4,

---

[55] The idea of the third circulation was introduced by Mirosław Pęczak, see his "O wybranych formach komunikowania alternatywnego w Polsce," *Kultura i Społeczeństwo* 32, no. 3 (1988); also his "Kilka uwag o trzech obiegach," *Więź* 31, no. 2 (1988).

[56] See Filip Musiał, "Dyskusja o końcu PRL," *Horyzonty Polityki* 5, no. 11 (2014).

1994, when the Act on Copyright and Neighbouring Rights was passed, the first regulation in Polish law referring to new media: video cassettes, satellite broadcasts, and computer programmes. But legal actions seldom bring quick, effective and total solutions to the problems they were introduced to solve. For these reasons, we decided to focus on the given medium's real diffusion and use it as the final point in the timescale of our description.

The concept of innovation diffusion was created by Everett M. Rogers in the early 60s and has been developed since then, becoming the basis for many detailed analyses.[57]

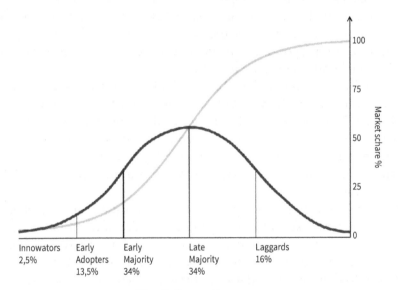

Fig. 2. Innovation curve

Source: Everett M. Rogers, *Diffusion of Innovations* (New York: Free Press, 2003), 273, 281

According to Rogers, all innovations follow similar patterns as they spread through a society. This refers to any kind of innovation: technical, social, or other. Innovations are first caught by innovators who are entrepreneurial, have a high social and financial status, and are willing to take the risk connected with the innovation. Innovators follow novelties in domains that they find interesting and often are competent enough to evaluate their worth.

---

[57] The book by Rogers was first published in 1962: Everett M. Rogers, *Diffusion of Innovations* (New York: Free Press of Glencoe, 1962). The newest, modified and extended version was published in 2003 (5th ed., New York: Free Press, 2003). It includes a critical analysis of his ideas.

However, they can sometimes misjudge their chances – after all, not all innovations catch on.

After the innovators come their followers, also known as trendsetters, who make up the larger group. They do not have to be fascinated with novelties nor know the technical details, but they, too, usually have a high social status. Their social position makes them cautious in their choices and less prone to take risks which could negatively impact this position. They are also more connected to local communities (Rogers calls them *localites* as opposed to cosmopolitan innovators) and they often play the role of opinion leaders.[58]

Later, the innovations are adopted by most of society, first by the "early majority" with their high social and financial status, then later by the more sceptical "late majority." In the end come the marauders, who are described by Rogers as the most traditional, poorest, and oldest, who are living their lives surrounded mainly by their families and close friends.

We need to remember that Rogers' deliberations are statistical in nature, and specific cases do not necessarily have to fit nicely into the groups he described. The model was developed to describe the free market situations, where innovations are not hindered by any insurmountable obstacles. In communist Poland, however, there were legal regulations that impeded the diffusion of innovations or made certain groups of users more privileged in accessing them.

The model by Rogers does not treat innovation as static, or as a certain ready whole which is simply spread among people. The idea of re-invention describes the change that an innovation undergoes when it is adopted and used. It is, then, mostly a process of negotiation which allows the users to react to the innovation and adjust it to their needs, and it is in this area where extra-systemic practices arise. Most of them are marginal in character, but we have found an interesting correlation. New media was adopted in Poland twice and in a twofold manner: the first time they were treated collectively, whereas the second time they were seen as individual home media. Microcomputers were available in community centres and schools, where many people made their first contact with them. Similarly, video technologies made their way to individual users' homes first through group screenings ("cassette cinemas"), and even satellite television was initially presented during mass events. It is not about the process – described by Rogers – of getting oneself acquainted with the innovation, at the same time as when we learn about its existence.[59] All these media were assimilated and functioned differently in Poland than in the West for some time, albeit each of them in its

---

[58] See Everet M. Rogers, *Diffusion of Innovations*, 5th ed. (New York: Free Press, 2003), 283.

[59] Ibid., 172 et seq.

specific manner. The first diffusion can be said to be an extra-systemic practice in terms of social protocols governing the use of media, as well as the political and legal system of the Polish People's Republic. At the same time, due to its universality (especially regarding microcomputers and VHS players), it can be seen as a norm founding such protocols. This was probably the opinion of those organising screenings, but also state planners preparing the openings of new state-owned videoclubs.

This double diffusion of new media resulted from a series of events. The first was Poland's technological delay, widely known and discussed. For example, the production plan for the MTV-200 VHS player was commented upon in such terms: "Unfortunately the Polish device, even at the beginning of its production, was not the most sophisticated design, it was mid-range, similar to designs already built in the West for years. In the year 1990, when it finally goes into production, it will already be obsolete!"[60] Similar opinions appeared every day, even in the technical press. An editorial in a 1986 *Audio-Video* magazine warned that the "Already much delayed Polish electronics face especially daunting challenges."[61] The people directly involved in the production of audio and video devices were fully aware of this backwardness. Jerzy Wojtas, who worked in Łódzkie Zakłady Radiowe Fonica (Fonica Radio Equipment Factory in Łódź) from 1960 until the early 80s and was then the Chief Expert on Research and Development, commented: "Everything that was manufactured in Poland in radio electronics was not impressive; it did not give the impression of being modern. I'm talking about the 60s here, but later we also felt this backwardness."[62] Everybody then, from engineers to ordinary users, felt that most or all advanced electronics were more likely imported than domestic. They had symbolic value, but also – slightly paradoxically – carried a certain amount of freedom from protocols regulating the use of these devices in the West (which sometimes resulted from a lack of awareness of these protocols or an inability to apply them). In the West they were subject to the laws of free-market economy, whereas in Poland one of the main goals of the *Audio-Video* magazine was to teach its readers how to make a DIY Cobra computer, together with a DIY keyboard. At the same time, the government was planning to create 2,000 village and neighbourhood videoclubs, where VHS movies could be watched.[63] This painful delay encouraged, or even forced, a wide range of extra-systemic practices.

---

[60] Ewa Czeszejko-Sochacka, "Rynek wideo: ruch w interesie," *Kino*, April, 1987: 25.

[61] Wiesław Marciniak, "Mikroelektronika – sprostać wyzwaniu," *Audio-Video*, no. 3 (1986): 1.

[62] Jerzy Wojtas, interview, December 20, 2016.

[63] Czeszejko-Sochacka, "Rynek wideo: ruch w interesie," 25.

The technological backwardness clashed with the discourse of progress. Modernisation programmes, which had been introduced in Poland since the 1920s, gained a Marxist-Leninist foundation after World War II, which saw technological progress as a historical necessity. It resulted in more attempts at modernisation, which were often incompetent and ineffective, but were nevertheless relentlessly undertaken. After industrialisation came the wave of computerisation and electronification, which resulted in a great number of engineers, and the popularisation of computers and electronics.[64] For many, these domains became a hobby, and were commonly associated with progress. When juxtaposed with the real backwardness it showed the flaws of left-wing ideology, which gave hope but failed to deliver.

The third circumstance was the rise of consumerism in Poland, advocated for by Edward Gierek, the First Secretary of the ruling Polska Zjednoczona Partia Robotnicza (PZPR, Polish United Workers' Party) from 1970 to 1980, who believed that "economic growth and large consumption will give the party social support and trust."[65] Among the consumed goods – apart from the usual food and clothing – there emerged a new group, namely household goods and consumer electronics. They were not associated with the old attributes of wealth, like gold or furs (which still functioned as a means for hoarding one's riches), but they were associated with technology and progress, which was highly valued in the communist ideology.

That combination of consumerism, technophilia and the feeling of backwardness, when compared with the western world, was characteristic of the 80s in the Polish People's Republic. For instance, in the book *Kasety magnetofonowe i magnetowidowe (Audio and Video Cassettes)*, which contained mostly technical information about magnetic recording, readers were advised:

> The list of global leading manufacturers of video tapes includes (in alphabetical order): AGFA, BASF, FUJI, JVC, Kodak, Maxell, Memorex, Panasonic, Scotch, and TDK. [...] It should be noted that Polish manufacturer STILON makes good video tapes for VCRs MTV-20 and MTV-50, type VC-30, VC-45 and

---

[64] In December 1961 the Economic Committee of the Council of Ministers passed the Resolution no. 400/61 on securing the conditions of manufacturing and using digital machines. In 1964 the position of the Government Plenipotentiary for Electronic Calculation Technology was created, which was replaced with the National Bureau of Informatics in 1971.

[65] Andrzej Leon Sowa, *Historia polityczna Polski 1944–1991* (Kraków: Wydawnictwo Literackie, 2011), mobi ebook. Edward Gierek opened communist Poland to Western consumerism but his economic policies based on foreign loans ultimately failed and led to the creation of the Solidarity free trade union movement.

VC-60. The tapes for these cassettes are imported from the best companies, such as BASF, AGFA, Memorex, Scotch, and ORWO, while the cassettes are produced locally.[66]

This advice created a certain attitude among consumers, and the expressed approval for the Stilon cassettes is ambiguous – it highlights the imported tapes and refers to cassettes for systems that were already obsolete.

Fig. 3. A cover of the first issue of *Audio-Video* magazine with an advertisement of a Philips "discophone": technophilia, consumerism and the feeling of backwardness

Source: *Audio-Video*, no. 1 (1984): cover

---

[66] Barbara Iwanicka, Edmund Koprowski, *Kasety magnetofonowe i magnetowidowe* (Warszawa: Wydawnictwa Komunikacji i Łączności, 1988), 174.

Methodology

In this book, the diffusion of innovation is more of a general framework rather than a research method. Our description of the spread of new media in Poland is not quantitative: we lack basic data, especially market information, to research the early adoption of new media.[67] Instead, we tried to describe the real ways media were used, especially those that were not entrenched in stable social protocols. We have assumed that such practices will be most common in the first stages of innovation diffusion, among innovators and early followers. Accordingly, we tried to find such people or the documentation of their actions. Usually, however, we reversed the usual method of historians who rely more on documents than verbal testimonies. Extra-systemic practices are sometimes illegal and often odd or humorous. Quite often they leave no trace in documents, and when they do, the image is distorted. Krzysztof Stasiak's recordings were not documented (even though the tapes themselves survived). Similarly, numerous video screenings, importing of equipment, computer clubs and other similar activities were also not documented. Even if there are some documents (including programmes or press articles from the time), they were often retouched or purposefully distorted: for instance, it was common to screen different movies to those that were originally announced. In such cases it is only through witnesses and organisers that we were able to investigate how new media operated at the time. For this reason, we have found the methods of oral communication and recordings of conversations with new media users crucial to our work and we based this book on these sources.

The methods of oral history have disadvantages and limitations, and many modern historians approach them with scepticism.[68] Because of this, it is easier to admit in advance that this book presents a media studies description and not a strictly historical account of events. We may also treat the information as a resource on cultural history, which has slightly less strict rules for validation than those expected by serious historians. But even the field of oral history presented us with a problem, namely the selectivity of interviews.

Reaching certain people proved difficult or untenable, and so it is not impossible that certain phenomena were overshadowed by other, more conspicuous

---

[67] A classic analysis of video recorder diffusion on the American market can be found in Bruce C. Klopfenstein's *The Diffusion of the VCR in the United States*, in *The VCR Age. Home Video and Mass Communication*, ed. Mark R. Levy (Newbury Park: Sage Publications, 1989). It is based on statistical data from multiple sources. This kind of research was not conducted in Poland and such information is, therefore, inaccessible.

[68] The discussion has been going on for years, arguments in defense of oral history were collected by Paul Thompson. See Paul Thompson, *The Voice of the Past. Oral History* (Oxford: Oxford University Press, 2000).

ones. For instance, it was difficult to encourage priests who had organised video screenings to recount these events and in most cases they would not facilitate us. It was also difficult to contact early users of new media in small towns, as well as to reach the women who participated in the new media transformation. We did not attempt to keep gender quotas, especially because in the new media culture of the 80s women were usually less involved than men. Moreover, with time, they seemed more reluctant to share their memories. Sometimes we were not able to find important contributors simply because they changed their names after getting married.

To sum up, we tried to describe the process of new media diffusion in detail, which may, in some cases, result in the lack of a broad perspective provided by quantitative data. It is the price we were willing to pay, if, in return, we could preserve some of the entrepreneurial, innovative spirit of that time.

Translation: Wojciech Szymański and Anna Czyżewska-Felczak

Bibliography

Auerbach, Jerzy. "Od redakcji." *Audio-Video*, no. 1 (1984).

Big's Gaming, "Audio Mastering – VHS Worth It?" Accessed January 2, 2017. https://www.youtube.com/watch?v=7iNhzEpmz94

Caillois, Roger. *Obliques*. Paris: Gallimard, 1987.

Carey, James W. *Communication as Culture. Essays on Media and Society*. New York: Routledge, 2009.

Castells, Manuel. *The Internet Galaxy*. Oxford: Oxford University Press, 2001.

Castells, Manuel. *The Rise of the Network Society*, 2nd ed. Malden, Oxford: Wiley-Blackwell, 2010.

Catholic Church. *Catechism of the Catholic Church*, 2nd ed. Vatican: Libreria Editrice Vaticana, 2012.

Chojnacki, Wojciech, Marek Jastrzębski. *Bibliografia publikacji podziemnych w Polsce. Tom drugi: 1 I 1986 – 31 XII 1987*. Warszawa: Editions Spotkania, n.d.

Cobb Kreisberg, Jennifer. "A Globe, Clothing Itself with a Brain." *Wired* 1995. Accessed December 12, 2016. https://www.wired.com/1995/06/teilhard

Cubitt, Sean. "Media Studies and New Media Studies." In *A Companion to New Media Dynamics*, edited by John Hartley, Jean Burgess, and Axel Bruns. Malden: Wiley-Blackwell, 2013.

Czeszejko-Sochacka, Ewa. "Rynek wideo: ruch w interesie." *Kino*, April, 1987.

Dewey, John. *Democracy and education*. Project Gutenberg. Accessed May 17, 2019. https://www.gutenberg.org/files/852/852-h/852-h.htm

Ellul, Jacques. *Technological Society*. Translated by John Wilkinson. New York: Knopf, 1964.

Feenberg, Andrew. *Questioning Technology*. London: Routledge, 1999.

Gajlewicz, Michał. "Nowe media." *Przekazy i Opinie*, no. 41–42 (1985).

Gitelman, Lisa. *Always Already New*. Cambridge MA: The MIT Press, 2006.

Gitelman, Lisa, and Geoffrey B. Pingree. *New Media, 1740–1915*. Cambridge, MA: The MIT Press, 2003.

Goban-Klas, Tomasz. "Nowe media – stare i nowe dylematy." *Przekazy i Opinie*, no. 51–52 (1988).

Greatbear audio & video digitising, "Digitising Stereo Master Hi-Fi VHS Audio Recordings." Accessed January 2, 2017. http://www.thegreatbear.net/audio-tape/digitising-vhs-hifi-audio-recordings

Heidegger, Martin. *The Question Concerning Technology And Other Essays*. Translated by William Lovitt. New York: Garland Publishing, 1977.

Iwanicka, Barbara, Edmund Koprowski. *Kasety magnetofonowe i magnetowidowe*. Warszawa: Wydawnictwa Komunikacji i Łączności, 1988.

Jakubowicz, Karol. "Nowe techniki informacyjno-komunikacyjne: czy mogą stworzyć nowy układ kultury." *Przekazy i Opinie*, no. 51–52 (1988).

Jenkins, Henry. *Convergence Culture: Where Old and New Media Collide*. New York: New York University Press, 2006.

Kamińska, Józefa. *Bibliografia publikacji podziemnych w Polsce, 13 XII 1981–VI 1986*. Paris: Editions Spotkania, 1988.

Keen, Andrew. *The Cult of the Amateur: How Today's Internet is Killing Our Culture*. New York: Doubleday/Currency, 2007.

Klopfenstein, Bruce C. "The Diffusion of the VCR in the United States." In *The VCR Age. Home Video and Mass Communication*, edited by Mark R. Levy. Newbury Park: Sage Publications, 1989.

Kołakowski, Leszek. *Main Currents of Marxism. Its Rise, Growth and Dissolution, Volume I: The Founders*. Translated by P.S. Falla. Oxford: Clarendon Press, 1978.

Kossak, Jerzy. "Innowacje elektroniczne i hipotezy zmian kulturalnych." *Przekazy i Opinie*, no. 1 (1975).

Liu, Yuping, and L. J. Shrum. "What is Interactivity and is it Always Such a Good Thing? Implications of Definition, Person, and Situation for the Influence of Interactivity on Advertising Effectiveness." *Journal of Advertising* 31, no. 4 (2002).

Lobato, Ramon. *Shadow Economies of Cinema. Mapping Informal Film Distribution*. London: BFI and Palgrave Macmillan, 2012.

Manovich, Lev. *The Language of New Media*. Cambridge, MA: The MIT Press, 2001.

Marciniak, Wiesław. "Mikroelektronika – sprostać wyzwaniu." *Audio-Video*, no. 3 (1986).

McLuhan, Marshall. *Understanding Media*. New York: Signet Books, 1964.

Musiał, Filip. "Dyskusja o końcu PRL." *Horyzonty Polityki* 5, no. 11 (2014).

Noll, A. Michael. "Videotex: Anatomy of a Failure." *Information & Management* 9, no. 2 (September 1985).

Pęczak, Mirosław. "Kilka uwag o trzech obiegach." *Więź* 31, no. 2 (1988).

Pęczak, Mirosław. "O wybranych formach komunikowania alternatywnego w Polsce." *Kultura i Społeczeństwo* 32, no. 3 (1988).

Pinch, Trevor J., and Wiebe E. Bijker. "The Social Construction of Facts and Artifacts: Or How the Sociology of Science and the Sociology of Technology Might Benefit Each Other." In *The Social Construction of Technological Systems*, edited by

Wiebe E. Bijker, Thomas P. Hughes and Trevor Pinch. Cambridge, MA: The MIT Press, 2012.

Pontifical Council for Justice and Peace. *Compendium of the Social Doctrine of the Church*. Accessed January 6, 2017. https://www.vatican.va/roman_curia/pontifical_councils/justpeace/documents/rc_pc_justpeace_doc_20060526_compendio-dott-soc_en.html

ProQuest. Accessed January 15, 2017. http://www.proquest.com/products-services/ProQuest-Dialog.html

Prószyńska, Krystyna, and Marek Błachut. "Teletekst." *Radioelektronik* 123 (1989).

Rheingold, Howard. *Virtual Reality*. London: Mandarin, 1991.

Rice, Ronald E. "Development of New Media Research." In Ronald E. Rice & Associates, *The New Media. Communication, Research and Technology*. Beverly Hills: Sage Publications, 1984.

Rice, Ronald E. "New Media Technology: Growth and Integration." In Ronald E. Rice & Associates, *The New Media. Communication, Research and Technology*. Beverly Hills: Sage Publications, 1984.

Rice, Ronald E., and Frederick Williams. "Theories Old and New: The Study of New Media." In Ronald E. Rice & Associates, *The New Media. Communication, Research and Technology*. Beverly Hills: Sage Publications, 1984.

Rogers, Everett M. *Communication Technology. The New Media in Society*. New York: The Free Press, 1986.

Rogers, Everett M. *Diffusion of Innovations*, 5th. ed. New York: Free Press, 2003.

Rogers, Everett M. *Diffusion of Innovations*. New York: Free Press of Glencoe, 1962.

Roszkowski, Jerzy. "Dyskusja nad sprawą 'nowych mediów.'" *Przekazy i Opinie*, no. 22 (1980).

Roth, Regina. "Marx on technical change in the critical edition." *The European Journal of the History of Economic Thought* 17, no. 5 (2010).

Sowa, Andrzej Leon. *Historia polityczna Polski 1944–1991*. Kraków: Wydawnictwo Literackie, 2011. mobi ebook.

Teilhard de Chardin, Pierre. *The Future of Man*. Translated by Norman Denny. New York: Image Books/Doubleday, 2004.

Thompson, Paul. *The Voice of the Past. Oral History*. Oxford, New York: Oxford University Press, 2000.

Vatican Council. *Pastoral constitution on the Church in the modern world: Gaudium et spes; promulgated by His Holiness Pope Paul VI on December 7, 1965*. Boston: Pauline Books & Media, 1998.

Wikipedia. "Teletekst." Accessed January 20, 2017. https://pl.wikipedia.org/wiki/Teletekst

Żebrowska, Halina. "Nowości techniczne." *Przekazy i Opinie*, no. 2 (1975).

Interviews

Krzysztof Stasiak, interview by Piotr Sitarski, Łódź, February 26, 2016.

Jerzy Wojtas, interview by Piotr Sitarski, Łódź, December 20, 2016.

PART II

PIOTR SITARSKI

VIDEO IN THE POLISH PEOPLE'S REPUBLIC: TECHNOLOGY AND ITS USERS

The beginnings of video technology

At the end of the nineteenth century, photosensitive tape was used to record motion pictures. It might have seemed then that this method solved all problems and satisfied the needs of everyone, especially as it introduced improvements that made it possible to achieve very durable copies of great quality. It is still possible today to watch films made a hundred years ago, and a 35 mm frame can have a resolution corresponding to 12 million pixels. Photosensitive tape has a disadvantage, though: we need to wait a while for the image to be developed. For that reason, when electromagnetically broadcast television appeared in the 1930s, people needed to find a way of joining its immediacy with the durability of the movie tape. The first television shows were just broadcast, and they were not recorded in any way. However, methods for recording the broadcast images appeared soon after. One of them, rooted in cinematography, involved filming the studio action, processing and developing the tape, and running it through telecine machine. This method was used by the BBC in the 1930s.[1] Another, more effective and cheaper method was developed roughly at the same time. It was called telerecording (kinescoping) and involved recording the images displayed on the screen on a movie tape. It was widely adopted and meant that it was no longer necessary to broadcast all shows live. It also solved the problem of broadcasting shows at the same time in different time zones, which was important, especially in the USA.

At the same time, research was conducted on recording images onto a magnetic medium, which was used since the end of the nineteenth century (Valdemar Poulsen's telegraphone) for audio recording. The most advanced laboratories were in Germany, and after World War II, German patents for magnetic tape production were obtained by the Allies. In 1956, the company Ampex presented the first device for TV image recording on a 50.8 mm magnetic tape, with transverse (scanning the tape across its width), and with four moving heads (hence the name quadruplex).[2] It uses a moving head rotating

---

[1] See Tony Currie, *A Concise History of British Television, 1930–2000*, 2nd ed. (Tiverton: Kelly Publications, 2004), 12.

[2] See John C. Mallinson, "The Ampex Quadruplex Recorders," in *Magnetic Recording. The First 100 Years*, eds. Eric D. Daniel, C. Denis Mee, Mark H. Clark (New York: IEEE Press, 1999).

in the direction opposite to the tape movement, which enables much greater speed of recording and, therefore, much longer recording time. In 1959, Toshiba developed the first method of helical scanning.[3] In their device, the head is tilted at a certain angle to the tape movement, which enables the track to be diagonally recorded and, thus, elongated.

This mechanism was quickly adopted in professional magnetic players; when compared to Ampex quadruplex technology, it was simpler, cheaper, and allowed for longer recording time. Moreover, it allowed for the still frame and fast or slow motion effects. These possibilities were especially important in sport broadcasts. Owing to these improvements, TV magnetic players became standard equipment in TV studios all over the world.

Their relatively low price also made it possible to use them in CCTV and home video systems. The first consumer magnetic players appeared in the mid-60s; they used helical scanning and 25.4 mm or 12.7 mm tape (inch and half-inch wide). The latter width was later used by the VHS system. In practice, different systems also used: ¾ inch, ¼ inch, 8 mm tapes, and others.[4]

Similar issues had arisen in other areas of consumer electronics, and earlier in cinematography. The main logistic problem was the necessity to establish common standards facilitating recording and playing tapes in different devices. Meanwhile, from the 60s until the 80s, numerous video systems appeared using a variety of tapes and slightly different ways of recording and playing them. Most of them were only locally significant, and initially, just tape was used for recording. The first magnetic player in Poland – the MTV-10, manufactured in Zakłady Radiowe im. Marcina Kasprzaka (Marcin Kasprzak Radio Equipment Factory) in Warsaw since 1973 – worked this way.[5] It was based on a model produced by Philips from 1966 (the LDL-1001) and recorded only black-and-white images.[6] Inaccuracies in the positioning of recording heads hindered the playback of tapes recorded on different devices, so the exchange of tapes among users was practically impossible. In the 70s, it became clear that devices which used tapes

---

[3] See Hiroshi Sugaya, "Helican-Scan Recorders for Broadcasting," in *Magnetic Recording. The First 100 Years*, eds. Eric D. Daniel, C. Denis Mee, Mark H. Clark (New York: IEEE Press, 1999).

[4] See Hiroshi Sugaya, "Consumer Video Recorders," in *Magnetic Recording. The First 100 Years*, eds. Eric D. Daniel, C. Denis Mee, Mark H. Clark (New York: IEEE Press, 1999).

[5] Studio television recorders were produced in Poland in the 1960s. See Roman Wajdowicz, *Historia magnetycznego zapisu obrazów* (Wrocław: Ossolineum, 1972), 153 et seq.

[6] See Unitra-klub, "ZRK MTV-10," accessed April 12, 2019, http://unitraklub.pl/node/1365

enclosed in cassettes worked much better in a home setting. In audio recording, cassettes developed by Philips had already been on the market for a decade; even though reel-to-reel recorders were still being manufactured, the future belonged to the cassettes.

After the MTV-10 was withdrawn from the market in 1975, it was replaced by the MTV-20 model, which could record colour images using Philips VCR cassettes. Another magnetic player from Kasprzak Factory was the MTV-50, which used VCR-LP cassettes, a modified version of the previous format, with longer recording time. These VCRs were home appliances, but it seems that they were seldom used in homes. In addition to reel and cassette recorders, the Kasprzak Radio Factory's catalogue from the first half of the 70s presents the first two models of video recorders. They were illustrated with photos depicting their professional (or, more often, semi-professional) applications: a folk band filmed by a huge television camera, and a young couple clearly enacting some theatrical scene outdoors. This contrasts with other pictures in the catalogue, where relaxed looking models posed with tape recorders in a conventional home environment. The cassette recorder was photographed with a child operating it, likely to suggest the simplicity of its use.[7] Juxtaposing home electronic devices with young, beautiful women was then a common practice, used also, for example, for the promotion of record players or radio receivers, bringing association of youth, modernity and carelessness, considered ideal conditions for joyful consumption. Apparently, however, VCRs were not to be marketed in this way. Several years passed before a similar advertising aesthetic was applied, and with it, brought similar consumer associations.

The purpose of these early appliances was semi-professional, rather than domestic. Of course, they did not provide enough quality for television, but they were used by the National Higher School of Film, Television and Theatre in Łódź in the newly created television studio in the city's Marysin district. Maciej Karwas, who ran the studio, recalls that professional equipment for recording television broadcasts was too expensive at the time for the School. Fortunately, one of the professors, and also the chief engineer of the Łódź and Warsaw television centres, managed to obtain a Philips video tape recorder (reel-to-reel type) through his personal contacts: "Jurek Ostrowski was, I think, visiting some event in France, like MIDEM, and while there he contacted Philips representatives. They gifted him with one of the first such models for amateur use."[8] Karwas recalls that this gift, when handed over to the Film School, was useful but problematic. Because it was brought without

---

[7] The catalogue is available at http://neurobot.art.pl/03/n-files/unitra/unitra. html, accessed June 27, 2018.
[8] Maciej Karwas, interview, May 15, 2014.

any documentation, the accounting department had problems with registering it. "For over three years until it broke down, the player was listed as inventory surplus."[9] Fortunately, it was replaced by the Polish video recorder:

> Meanwhile, the Warsaw Kasprzak Factory started producing a very similar Polish player. They were called MTV-10, which also used open tapes. We got three, maybe four such players from Kasprzak and could record those etudes of ours, but at rather poor resolution, because it was 200, 250 lines, so below the standard of a decent VHS, not to mention Super VHS, which had much better recording parameters. But you could demonstrate how it worked, you could discuss all the shortcomings with the students; production, staging glitches, camera movement – all that was possible. And later came the MTV-20, which also used square cassettes, where a roller was placed over a roller and the tape went from the upper one to the lower. Licensed by Philips, they recorded video at a slightly higher resolution.[10]

It is difficult to say whether the issue of lack of adequate documentation for the Philips recorder was a serious problem, and if it had any consequences. It might have been just an amusing anecdote. It seems at least symbolic to me, however, that introducing a video recorder into an institution caused such a confusion. This initial "extra-systemic" status of the Philips recorder in Łódź Film School foreshadowed the later cultural history of VCRs, the uncertainty regarding their status as professional or amateur devices, and the economic challenges that the VCRs caused.

In the 1970s a group of students and graduates from the National Film School formed the Workshop of the Film Form, formally a research circle, which soon became an avant-garde group exploring structural cinema.[11] Workshop participants were interested in television and video technology and were frequent visitors to the School studio in Marysin. Of course, the achievements of the Workshop cannot be reduced to video technology, but the video and television technological background of the Film School played a crucial role in creating movies such as *Transmisja przestrzenna* by Wojciech Bruszewski (1973) or the *Video* cycle by Paweł Kwiek (1974–1975). Thus, video technology inaugurated an important period in Polish avant-garde art. Józef Robakowski, a member of the Workshop and an acknowledged video artist, juxtaposed the Workshop's approach with television, pointing to the peculiarity of video art as opposed to the mass character of television in one of his manifestos: "Video art is opposition

---

[9] Ibid.

[10] Ibid.

[11] See Marika Kuźmicz and Łukasz Ronduda, eds., *Warsztat Formy Filmowej / Workshop of Film Form* (Warszawa: Arton Foundation / Fundacja Arton; Berlin: Sternberg Press, 2017).

depreciating the functional nature of this institution [television – PS]; it's an artistic movement which, through its independence, exposes the mechanism of controlling people, putting pressure on them and suggesting to them how to live their lives."[12] Robakowski pointed out that, although in the technical aspect video was identical to television, video has the advantage of offering completely new tools for artistic communication. He put this claim into practice in his later movies, especially during the 80s, with films such as *Pamięci L. Breżniewa* (1982) and *Sztuka to potęga!* (1984). Robakowski's statement can be seen in even more general terms if we accept that the new possibilities of expression were offered not only by video cameras and were not limited to artists: video technology, as a medium, offered its users completely new ways of using it.

MTV-50 recorders were also used – apart from training and artistic purposes – for organising public video screenings. Jacek Rodek, a student activist and one of the Polish sci-fi fandom organisers (and then later an editor and publisher), recalls an example: "I met a man working in the now closed Unitra factory, which made the first Polish video recorders. It wasn't the kind of equipment you could buy in a shop, but, like everything at the time, we could get it from under the counter. And so I did, investing my own money."[13] Rodek specifies in the interview that the recorder required Philips VCR cassettes.[14] He later used the same device to organise video screenings in student clubs[15] in Warsaw and other cities.

The video recorder was not an appliance one could just buy; it had to be procured. Although any use of this device was, in this case, institutional (as it was used in student clubs), it was owned personally by Jacek Rodek, who emphasised this facet several times during the interview. He stressed that screenings were very popular, but any money made went to the clubs and not the organisers. It is obvious that video recorders escaped the contemporary norms: they were not strictly private, like TV sets, but neither were they owned only by institutions, like cinema projectors. The same is true not only for the late 70s. A similar, even more intricate story is told by Robert J. Szmidt, another organiser of early video screenings who later became a science fiction writer and publisher, and another prominent figure in Polish sci-fi fandom. In 1982 he bought a VHS player and started

---

[12] Józef Robakowski, "Video Art – Szansa podejścia rzeczywistości," *Gazeta Szkolna PWSFTViT*, 1976, accessed January 2, 2017, http://repozytorium.fundacjaarton.pl/index.php?action=view/object&objid=3173&colid=75&catid=18&lang=pl

[13] Jacek Rodek, "Seks, kłamstwa i kasety wideo," in *Hybrydy. Zawsze piękni, zawsze dwudziestoletni*, ed. Sławomir Rogowski (Warszawa: Fundacja Universitatis Varsoviensis, 2013), 114.

[14] Jacek Rodek, interview, February 4, 2015.

[15] Student clubs were centres of cultural activity such as music concerts, film screenings, or theatre performances. They were run by students' associations.

collecting movie cassettes, before later agreeing to sell the device at the request of the School Council of the Polish Students' Association at Wrocław University of Science and Technology. The association hired him as its operator, as he was the owner of the cassettes. It did not change the way the player was used. Szmidt recollects: "I decided I would keep it with me and bring it just for the events."[16]

Early conceptualisations of video technology in Poland

When considering the situation of VCRs in the 1970s and 80s, we situate them in a historical process of technological development. Thus, they can be seen as an improvement on, or perhaps a crowning of, magnetic recording technology, i.e. "more advanced audio tape recorders." They can be also placed within and described as part of the process of cinematographic recording development, and we can see them as the beginning of magnetic image recording. Video tape recorders also belong to the history of television, being an extension of the TV set. Finally, they can be looked at from the perspective of user experience and seen as a stage in the development of home cinema systems. All these perspectives are important, but in the end, they obscure the real perceptions of their users at the time, and how the cultural interpretation changed from when they were something "obvious" at first, to when they went out of use – a dusty, bulky piece of junk.

In the popular film press, news about magnetic recording appeared for the first time in 1965. Gideon Bachmann wrote in his article in *Film* magazine, provocatively or humorously titled "Czy będziemy kupować filmy w sklepach?" ("Will We be Buying Films in Shops?"), about recording moving images on magnetic tape: "The development of electronic methods of recording and playing films will enable in the not-too-distant future a price drop of this currently very costly equipment, and in turn, people will be able to create small home cinemas."[17] At this early stage the terminology had not been established yet, the new equipment did not have its name, and the author did not have a clear vision of the development of the technology. He uses the metaphor of a home cinema, though, which will become a recurrent theme.

The topic of magnetic recording returns in 1970, when the magazine *Film* organised a debate, tellingly titled "Film on a cassette – cinema at home." This term – "film on a cassette" or just "cassette" – was used to describe all forms of video technology in Polish popular film magazines in the 70s. It was conveniently universal as disc systems, like TeD or CED, and used discs placed in plastic cassettes

---

[16] Robert J. Szmidt, interview, April 25, 2015.
[17] Gideon Bachmann, "Czy będziemy kupować filmy w sklepach," *Film*, May 2, 1965: 7.

or paper casings. Interestingly, because the participants of the debate apparently had not been personally acquainted with video equipment, they used the term "cassette" in reference to the whole device.

Fig. 1. The term "cassette" is used to describe the whole device for magnetic video recording

Source: *Film*, July 26, 1970: 6

General comments and predications from the participants of the debate were, surprisingly, almost prophetically accurate. They foresaw the two-stage diffusion of the new medium (first collective, then – with decreasing prices – individual), the creation of film rental shops, the revolution in distribution, the change in TV offer, and so on. Their attention was concentrated on the medium of the cassette, and the video players themselves seemed less important. The unclear status of the new technology is reflected by the lack of uniform terminology. When Wiesław Stempel, Head of the Technical Department of the Central Film Office (Naczelny Zarząd Kinematografii), referred to them, he chose the awkward term "recording-playing devices."[18] Later, he wrote about an "add-on to the TV set."[19] The term *magnetowid* (video cassette recorder), which later became standard, was seldom used in Polish press in the 70s. In his report from Rome, Virgilio Tosi recounted Italian debates on the new film technologies. In the first instance he mentions "video tape recorders, called here 'cinema in a cassette.'"[20] Generally, video was

---

[18] Wiesław Stempel, "Film w kasecie – co nowego? Mówi dyrektor Wiesław Stempel," interview by Elżbieta Smoleń-Wasilewska, *Film*, November 22, 1970: 11.

[19] Wiesław Stempel, "Pierwsze słowo techniki," *Film*, August 9, 1970: 6. The term "add-on" was used in a general sense but the author later concentrated on EVR technology.

[20] Virgilio Tosi, "Kiedy umrze kino," *Film*, January 11, 1970: 12.

seldom discussed in the Polish film press in the 70s. When it was mentioned, usually in the international news sections, attention was concentrated on the possibility of owning cassettes and freely choosing the films. The cassette player itself seemed far less important.

Beside "films in cassettes," the term "home cinema" is often used and it requires a more detailed explanation. The transfer of cinematic experience into private space began when the first such experiences took place, namely at the beginning of the 20th century. Kazimierz Prószyński's "Oko," first presented in 1914, was both a video camera and a projector, and was a great tool for both making and watching amateur films. Prószyński did not manage to market his device, but from the beginning of the 20s the Pathé Baby 9.5 mm perforated tape system started to gain popularity. While Prószyński saw Oko primarily as a camera, Pathé Baby was, from the start, designed as a device for individual rental and playing of films distributed by the company Pathé Frères. The system was relatively popular in Poland.[21]

After World War II the practices of home cinema continued. Zakłady Kinotechniczne (Cinema Equipment Factory) Prexer in Łódź started producing the 8 mm projector Amator in the 50s, and, later, a second model, the AP-33 Polux. Apart from these two, there were also German and Soviet devices available in Poland. Far more popular than 8 mm projectors were slide projectors and filmstrip projectors. Slides and filmstrips were inexpensive. They featured stories for children, often based on films or TV series. Because they were accompanied by text which organised the visual material narratively, family filmstrip shows developed into elaborate imitations of cinema screenings.

The authors of articles in *Film* in the 1970s conceptualised films on cassettes as a continuation and improvement of these technologies. Reality, however, proved them wrong. The video tape recorders in the 70s and 80s offered insufficient image quality.[22] Connected to usually small, low-end TV sets, they could not match the power of expression of not only cinema but even that of home slide projectors. But they did not need the room to be dark nor any other special arrangements (such as setting up a screen or a white background). Using them was convenient, but devoid of the unique atmosphere of a cinema theatre. Additionally, the VCR format used in the MTV-20 model was limited to just 60 minutes, which made it impractical for watching full-length movies.

Under these circumstances, the other basic conceptualisation of early video tape recorders should not be surprising – they were perceived as television, not cinematic devices. In 1972, when *Film* wrote about "films on cassettes," the book

---

[21] Zbigniew Cybulski, a famous Polish actor, was amazed by Pathé Baby (see Jerzy Afanasjew, *Okno Zbyszka Cybulskiego* (Warszawa: Prószyński i S-ka, 2008), 27).

[22] Horizontal resolution of VHS was 240 lines.

Telewizja kasetowa (*Cassette Television*) by Bolesław Urbański was published. It described a number of contemporary methods of magnetic recording but presented them as a continuation of television: "by the term 'cassette television' we understand the reception of a television broadcast – recorded on films, tapes, or discs – by means of a TV set."[23] Despite the fact that conceptualising video technology as cassette television may now seem exotic, it permeates many press articles in the 70s. We should also remember that magnetic tape recording was then just one of several possible choices, and it was not clear which would finally surpass the others. Bolesław Urbański suggests that these systems could coexist symbiotically: "Private users will use TV reels, cassettes, and discs at home in a way similar to the way they do now with vinyl records and magnetic audio tapes."[24] Alternative systems using optical recordings, such as super 8 mm telecine or Electronic Video Recording (EVR), seemed attractive at the time. In the end, however, the former turned out to be too complex and the latter too expensive – even though the EVR system offered excellent quality when compared to later magnetic recording systems.

An interesting example of "cassette television" is the use of video recorders at the Henryk Arctowski Polish Antarctic Station. Jacek Siciński, professor of biology and participant in several Polish polar expeditions, recalls video players at the Station during the fourth expedition:

> I was at the station for the first time in 1979, and then [Philips] VCR video recorders prevailed there, that old system still. I think they were brought together with the cassettes when the station was being built… In 1976, two ships, "Dalmor" and "Zabrze," had already arrived with materials for the construction of the station. All this was there because the first group was staying there during the winter, so they needed these video players most then. Well, of course, apart from tape recorders, the reel-to-reel ones, everyone had them. I remember these reels very vividly, these cassettes, they were compact, unlike those of the VHS system which was introduced later.[25]

Siciński also gives an account of the content of the cassettes: "I remember mostly programmes that I used to like in my childhood, in my youth; cabarets and other such entertainment programmes. […] It was all poorly recorded… There were also cassettes with original films, I think."[26] He does not remember the details, given that years have passed; it is possible that his memories about

[23] Bolesław Urbański, *Telewizja kasetowa* (Warszawa: Wydawnictwa Komunikacji i Łączności, 1972), 7.

[24] Ibid., 10.

[25] Jacek Siciński, interview, March 8, 2017.

[26] Ibid.

the originally recorded films come from later polar expeditions and VHS cassettes. Certainly, though, at least part of the content of VCR cassettes was television programmes. When describing the use of the video recorder, Siciński also emphasises that it was "fast entertainment," which he clearly contrasts with film screenings organised with two 16 mm projectors, which were more formal.

The use of video cassettes for recording and then playback of TV programmes had, however, limited attractiveness unless in special situations like the Antarctic station or on cruises (where video players were initially operated alongside traditional 16 mm projectors, and eventually displaced them). Essentially, however, viewers wanted to watch what TV did not broadcast.

The 70s also saw the development of disc recording systems. The first was the Television Electronic Disc format (TeD), created by German companies Telefunken and Teldec. The information was recorded in similar fashion to the gramophone, but the track pitch was reduced to 0.007 mm, making it possible to record 130-150 grooves per millimetre (compared to 10-13 grooves on an audio disc). The TeD flexible foil disc spun 1,500 rpm on a cushion of air, and a piezoielectric needle converted the vibrations into an electric signal. Eight-inch discs could store five minutes of programming, 12-inch discs about 12 minutes – hardly enough for a home cinema system.[27] However, this capacity was enough for short TV shows, usually nature documentaries, fragments of sport programmes, and animated films for children. In this way some Polish productions, most notably the adventures of Bolek and Lolek, which were popularised in West Germany by television, were available on Teldec discs.[28]

Roman Wajdowicz, an expert on video recording methods, and the rector of Łódź Film School, saw TeD discs as the medium that was going to dominate the market in the future: "The low cost of a flexible Teldec or TeD video disc makes it a perfect information medium, which, because of the speed of recording and copying, can be used as an addition to magazines or even daily newspapers."[29] However, this format, which was introduced in 1975, did not gain popularity. It was replaced by other formats of disc recording: grooveless capacitance discs (with a needle sliding the disc surface and reading the changes of electric capacity) and optical recording discs (using lasers).

---

[27] See Terramedia, "TeD video disc," accessed December 12, 2016, http://www.terramedia.co.uk/media/video/ted_video_disc.htm

[28] *Bolek and Lolek* (also known as *Bennie & Lennie*) is a Polish cartoon comedy series produced from 1962. The series proved to be extremely popular not only in Poland but also around the world, becoming one of the biggest export hits of the Polish film industry.

[29] Roman Wajdowicz, *Nowoczesne metody rejestracji obrazów* (Warszawa: Komitet do Spraw Radia i Telewizji, 1975), 69.

Fig. 2. The adventures of Bolek and Lolek recorded on a TeD disc

Source: The cover of a Telefunken-Decca Schalplatten GmbH record

Recording on discs had numerous advantages over tape recording. Firstly, discs enabled faster access to any part of the material. Secondly, they were based on a simple, cheap and reliable drive, continuously developed since the 1920s. These features finally cemented the technology and was later used in computer floppy discs and CD, DVD, and Blu-Ray disc drives. In the 1970s disc recording seemed the more obvious solution as a continuation of the well-known gramophone recording. These circumstances resulted in an attempt to construct a Polish video disc system, initiated at Łódzkie Zakłady Radiowe Fonica (Fonica Radio Equipment Factory in Łódź).

At the time, Fonica produced television components, measuring equipment and predominantly jukeboxes and gramophones. Using the same components, it also manufactured ventilation fans, which may seem irrelevant but proves how simple and reliable the spinning mechanism is, invented by the early agricultural civilisations as the potter's wheel and still employed in contemporary computer hard disk drives. Fonica's engineers followed the market of technical novelties and were interested in Telefunken's TeD project. "Our interest in the issue, the issue of the videodisc, boomed right after we saw a Bildplatte," recollected Jerzy Wojtas, who was the head of the Research and Development Office.[30] It seemed promising because at the time Fonica was collaborating with Telefunken, producing the gramophone G-500 under their licence. Unfortunately, Telefunken was

---

[30] Jerzy Wojtas, interview, December 20, 2016. My account of the Fonica video disc project is based on this interview.

unwilling to share their knowledge on the video disc. It soon transpired, however, that other methods of recording were better. As a result, the engineers at Fonica chose laser recording. Jerzy Wojtas organised a laser laboratory and hired an expert from the Łódź University of Technology to build a laser and deal with modulating its light. Wojtas admits that there was little hope for creating an independent design for a video disc recorder: "Was that supposed to lead to production? Nothing we did was supposed to lead to production. We were just learning."[31]

Meanwhile, the Consortium of Polish Consumer Electronics Manufacturers Unitra, which Fonica belonged to, started collaborating with American company RCA, to obtain a licence for producing television cathode ray tubes. Jerzy Bilip, a technical director at Unitra, received RCA's capacitance electronic discs during one of his visits to the USA. Bilip asked Wojtas if his Fonica research team could play them back. When, after several attempts, they did, Unitra acquired the licence; Wojtas remembers that it cost only $300. At the end of the 70s, the Fonica research team began to build a video disc player under that licence.

Independent manufacturing of discs proved impossible, however. The production technology was too complex for the Polskie Nagrania (Polish Records) pressing plant, the main producer of vinyl records in Poland. Fonica decided to produce discs in the USA and then distribute them both in Poland and abroad. Jerzy Wojtas prepared the test discs, recording a music concert, after the state monopolist, the Film Distribution Office (Centrala Wynajmu Filmów), refused the rights to its films. Wojtas recollects these preparations:

> I went to the Film Distribution Office and presented this problem, and the director of the Film Distribution Office said to me: "I'm sorry sir, but I don't know how to do it. I don't know, because I don't know who owns it. I have the right to send a film to cinemas and charge them for it. [...] And if somebody wants to take a film abroad to show it – off they go. But if a film is to be copied onto something, I don't know how to go about that.[32]

This anecdote shows, again, the inability of the legal system in the Polish People's Republic – this time at a very high level – to respond to the challenges of distributing films on disc or cassettes.[33]

Fonica's project aroused a lot of interest, especially when it was presented on popular TV science show *Sonda*. Unfortunately, the introduction of martial law in 1981 ended the collaboration with RCA and almost all work was halted

---

[31] Ibid.

[32] Ibid.

[33] The problem could have been solved since TeD discs with Polish TV programmes had been produced earlier.

on both the capacitance and laser technologies. When later the disc recording technology started dominating the market (CD, DVD), Fonica lacked both specialists and experience. The factory never manufactured video disc players but it cooperated with Japanese Mitsumi and Taiwanese Hanpin companies to produce CD-players in the late 1980s.[34]

Fig. 3. "Vision record? Videorecord? Videodisc?" – in search of a name for the video disc. The magazine also announced a contest to name the video disc player

Source: *RTV. Radio i Telewizja*, April 8, 1979, cover

The videotape format war and the cultural adaptation of the videocassette player

In the late 70s and early 80s, it seemed that the future of recording belonged to cassette tape recorders. Three most widespread formats competed at the time. The changes that were taking place were surprisingly aggressive – video technology was one of the first media where competition between formats got intense. Both of the aforementioned formats, VCR and VCR-LP, went out of use, and the

---

[34] See Paweł Cendrowicz, "Fonica w świecie CD, czyli 'dyskofony' z Łodzi," accessed November 7, 2018, http://www.technique.pl/mediawiki/index.php/Fonica_w_%C5%9B-wiecie_CD,_czyli_%E2%80%9Edyskofony%E2%80%9D_z_%C5%81odzi#Podsu-mowanie.2C_ocena_lub_rachunek_sumienia

so-called "war of the formats" was fought between three methods of recording (in chronological order): Betamax, VHS, and Video 2000. The first of them, developed by Sony, replaced its other format, U-Matic. It was introduced in 1971 but turned out to be too expensive, and the company decided to target it at professional markets. Betamax, which appeared in 1975, was designed to be used at home.[35] The "war" actually started in Japan when Sony failed to gain unanimous support from all domestic companies. JVC, a part of the Matsushita company, offered a competing recording format – VHS (Video Home System).[36] This competition had different impacts on different markets, and although the factors were the same (price of devices, recording length, audio and video quality), they were not equally important in the USA, Western Europe, and Japan. In the mid-70s, VHS proved to be the obvious winner. Philips finally abandoned its Video 2000 in 1983, and while Sony continued to produce Betamax devices their role in the leading world markets was marginal.[37]

The situation in Poland was peculiar and much less affected by the forces of free market economy (although they, legally or illegally, had some impact). Philips' Video-2000 system was quite popular, which resulted mainly from the fact that video recorders from that system were officially sold in the country, imported by the company Konsuprod.[38] Maciej Karwas recollects that this was his first video tape recorder: "I bought that Video-2000 tape recorder from a football player – it was Smolarek, a player from Łódź."[39] In the end, the success of VHS on the main global markets made it also spread to and dominate peripheral markets, including Poland. In the mid-80s, the media landscape was already stable. Paweł Gaweł, who described the video market in 1984, estimated that VHS had an 85% share, and the remaining systems shared the other 15%.[40] He saw the main reason for this in the popularity and affordability of the VHS system abroad. He also attributed the popularity of other formats at the very beginning of the 80s to buyers' ignorance:

---

[35] See Wikipedia, "Videotape Format War," accessed December 12, 2016, https://en.wikipedia.org/wiki/Videotape_format_war

[36] In 2008 the traditional name Matsushita, coming from the family name of its founders, was replaced by a new one – Panasonic.

[37] See Hiroshi Sugaya, *Consumer Video Recorders*, 189 et seq. See also Marc Wielage and Rod Woodcock, *The Rise and Fall of Beta*, accessed December 11, 2016, http://www.betainfoguide.net/RiseandFall.htm

[38] See Piotr Gaweł, "Rynek wideo w Polsce," *Film na Świecie*, no. 334–335 (1986): 53.

[39] Maciej Karwas, interview, May 15, 2014.

[40] Piotr Gaweł, "Zasięg video w Polsce," *Zeszyty Prasoznawcze* 25, no. 3 (1984): 165.

As highly luxurious goods, video players were initially bought by people who worked abroad and had sufficient income, such as doctors or other specialists. At the time, no system dominated on Western markets. Under these circumstances, being less market savvy, they decided to buy video tape recorders made by established manufacturers, that is Philips and Sony with their V-2000 and Betamax systems.[41]

The case of Maciej Karwas clearly contradicts this claim. While Włodzimierz Smolarek, a football player, could have been a wealthy layman, Karwas, who bought the V-2000 device from him, was as well-informed as he could be. Gaweł was also critical in his opinion about the backwardness of the Polish industry: "An example of wrong and inflexible choices regarding the range of products was the idea to manufacture around 2000 obsolete VCR devices a year in the Kasprzak factory in Warsaw, cassettes for which were no longer produced anywhere in the world."[42]

It is true that the process of decision making was rather slow in the Polish industry, but video tape recorders were one of the first devices which entered the market so dynamically while competing so fervently with other standards. This battle made it impossible to maintain the "controlled delay," which, more or less intentionally, was a part of the modernisation policy in the Polish People's Republic. It was based on optimising the costs of purchasing foreign licences and copyrights. The most modern technologies were unavailable due to legal policy (the CoCom, Committee for Multilateral Export Controls embargo on technologies that might have been potentially used for military purposes) or simply because they were too expensive. It was possible to control the delay as long as the technological progress was fast but did not involve the complete replacement of technical standards. For instance, TV sets and radio receivers aged slowly and kept their functionality. Video tape recorders brought about an increased pace of standardisation, in which just a few-year-old devices, technically perfectly serviceable, were completely obsolete because there were no new cassettes that could be played on them. The costs of the hard currency share, including, in this case, licenses and components imported from abroad, also increased, and as a result production became unprofitable.

In the first half of the 1980s the Polish electronics industry, as well as the entire country's economy, collapsed. The situation was further aggravated by international sanctions introduced after the declaration of martial law, which brought the suspension or termination of previously established foreign contacts. Jerzy Wojtas describes martial law in Fonica: "Collaboration with the Americans was denounced in every field. For, I don't remember how long, about a month, the phones didn't work. People came to work and did nothing; actually, everyone

---

[41] Ibid.
[42] Gaweł, "Rynek wideo," 53.

was only sitting around and talking."[43] Production continued, but the developmental work had to be dramatically stopped. The format war made the license tactics ineffective because it was not clear which licenses to buy. All these circumstances meant that in the second half of the 80s the Polish electronics industry was lagging even further behind than before. The VHS MTV-100 video recorders manufactured by Unitra since 1986 could not compete with foreign equipment imported in large numbers, and later Polish video recorders from Unitra (Bondstec BT 310, MTV-300, GoldStar GHV-1223K, Polkolor-Schneider SVC 265) were only assembled in Poland from foreign components and did not play a major role in the diffusion of the video technology in the country.[44]

At the beginning of the 80s, video tape recorders were still "strange" devices, not fully conceptualised nor tamed. The idea of "having" a film on cassette and watching it on a TV set seemed odd. Agnieszka Nieracka recalled the following story:

> I remember such a moment with perfect clarity. I was living in a dorm at the time, a hall of residence for university teachers, in Katowice. Someone came to see us and we played a joke on him. We covered this big device with a towel while a film was playing. I remember what it was: *Bring Me the Head of Alfredo Garcia*. And this friend of mine comes in, now a renowned professor, and says, 'Dear lord! That film is on television!' And off he went, to his own room, to watch it. It didn't occur to him that we could have such a device. My husband got it from somewhere. I think it was from the Physics Department, from the physicists.[45]

Anecdotes of this sort, even if coloured by the passing of time, illustrate that the device triggered a sense of wonder: it was unknown, existing somewhere beyond the sphere of familiar media. Agnieszka Nieracka remembers one more characteristic story:

> I remember something kind of metaphysical: that I was carrying a video cassette through the city. It was in a town called Sosnowiec, where I worked at the University of Silesia. After, we moved to Katowice, specifically to RTV [Radio and Television Department]. Anyway, I was carrying this cassette and the film I remember, was *Blade Runner*. And for me it was this amazing. That's why I'm talking about metaphysics, because, oh god, I had the film. There was a film inside. Maybe it wasn't *Blade Runner*? I really don't remember now.[46]

---

[43] Jerzy Wojtas, interview, December 20, 2016.

[44] For the description of Polish video recorders see Unitra-klub, "Magnetowidy," accessed April 12, 2019, https://unitraklub.pl/magnetowidy

[45] Agnieszka Nieracka, interview, December 16, 2016.

[46] Ibid.

It is very difficult to find out exactly when this took place. Cassettes with *Blade Runner* appeared in Poland around 1983 and the Department of Radio and Television of University of Silesia was founded in Katowice in 1978. The related anecdote could have taken place in 1977 (the University of Silesia surely had video players by then, and the movie could have been something other than *Blade Runner*), as likely as in 1984. No matter what the date was, this story confirms early fascination with the new medium, but also the uncertainty and amazement it caused.

In the mid-80s, when Paweł Gaweł published his papers on video, this fascination was gone. Gaweł looked at the medium from a "black box" perspective, when the final interpretation of the new technology had already taken place and other possible ways of understanding it were just an obsolete curiosity (with the exception of public video screenings which will be discussed later). He ignored the disadvantages of the VHS format: lower image quality when compared to its competitors and shorter length of recording than V-2000, which could contain up to 16 hours of recorded material. He also omitted all other possible applications of video technology and concentrated on what the video recorder finally became: a simple to use household appliance for playing films pre-recorded on cassettes.

Video technology underwent a long evolution, from being conceptualised as home cinema, as a collection of films on cassettes, and as cassette television. Then finally, in the mid-80s, it was interpreted as a completely new form of home entertainment for family and friends. It was not a home cinema, and it was different from 8 mm film screenings or slide shows because of its ease of use. It was almost fully automated, as it was enough to put the cassette in to watch a movie, and the buttons allowed users to pause, stop and rewind (which was often highlighted in advertising). Video tape recorders were also one of the first remote controlled home appliances. Their modern interactivity replaced the magic of a dark cinema theatre, and the relatively large capacity of video cassettes enabled recording movies and creating home collections, which ultimately separated video from television. Moreover, the repertoire did not depend – at least in Poland – on the supply, but rather, to a large extent, on logistics and the number of friends and acquaintances a person had. It is paradoxical that at least some of these characteristics of the medium of video were not actually experienced by Polish video users in the mid-80s – they were imagined: partly imported from the West, partly resulting from people's dreams. Nevertheless, the image was shared even by those who had contact with video players only through public screenings.

Cultural adoption of video technology in Poland relies on the image of VCRs as objects of consumerist fantasies and as the vanguard of Western affluence. The westernisation of Polish-made equipment is interesting in this respect. The production of video recorders in Poland has never been fully independent, and even the early models were based on Western solutions. Their appearance, however, suggested local origin. The buttons and knobs on the MTV 10 were described in Polish, and the name of the device on the casing ("magnetowid") was also Polish. The MTV 20 employed pictograms. Newer models, however, pretended to be imported equipment. MTV 100 was described on the casing in English as "Video Cassette Recorder," and all buttons had English names. Its successor, the MTV 300, was even described in the manual as a "Video Record-er." Whereas Polish cassettes (manufactured or actually packaged since 1988 in the Stilon Gorzów factory) had both Polish and English names printed on them, their "brand" and description of quality was in English only: "Super High Grade." Marketing considerations played a definite role here. English subtitles suggested that these goods were intended for export, and therefore their quality was higher. They also signalled Western luxury and technological advancement.

Videocassette players in the 80s: users and sources of origin

Both quantitative and qualitative data regarding the spread of VCRs is sub-ject to debate. In both cases, we are forced to rely on estimates and speculations because the diffusion took place mostly outside the official circulation and it is difficult to guess the number of devices imported to Poland outside of the Pewex and Baltona chains.[47] Piotr Gaweł, the only scholar who researched video during it emergence in Poland, estimated the number of VCRs at 3,000 in 1981, and 15,000 by the end of 1982. Different estimates for the year 1983 vary between 20,000 and 70,000, while the following year there were as many as 150,000 VCRs. The author estimated the number of devices at 400,000 at the end of 1985.[48] It is not possible to verify this data, as it was not based on any identifiable sources (or, at least, the author is not quoting any).

Tab. 1. VCR saturation per 100 units (in percentages)

Measure \ Year	1981	1982	1983	1984	1985	1986
Per 100 black and white TVs	0.04	0.18	0.82	1.7	4.5	6
Per 100 colour TVs	0.15	0.75	3.5	7.5	20	27.5
Per 100 households	0.025	0.13	0.58	1.25	3.33	4.6

Source: Piotr Gaweł, "Rynek wideo w Polsce," *Film na Świecie*, no. 334–335 (1986): 76

---

[47] Pewex and Baltona were chains of shops that accepted payment only in US dol-lars and other hard currencies.

[48] Gaweł, "Rynek wideo," 63–64.

Gaweł's article, published in the midst of the VCR revolution, is impressive because of the data the author was able to gather (or sometimes – to guess), and the methodological framework (although in a few places the author refers to his "own observations," not specifying their nature). At the same time, however, the paper presents a teleological vision of the medium's development, where from the very beginning video has an "essence": not fully developed at first, but gradually reaching maturity. The periodisation proposed by Gaweł corresponds with this. According to him, the first period began in 1980, when the possibilities of the new medium became available to the general public. The second period, between 1983 and 1984, was marked by initial fascination with the medium, while the third period (1985–1986) combined the fascination with the full demonstration and availability of the technology. The last phase, of complete maturity, was expected soon, when VHS recorders would become essential home appliances in most households.

Such an approach ignores the first years of consumer video technology diffusion (beginning in 1973, when the production of Polish video recorders began, although such devices had been present in Poland even earlier). Of course – there were incomparably fewer video players in the first decade of the medium's diffusion, and their social function was different, but this does not have to be seen as an "imperfect" form. Subsequently, the decade of the 80s is perceived as a period when video slowly enters into its last stage of development, i.e. home screenings and rental shops. In the following part of my analysis, I question this view and propose that the history of video, including the period of "cassette cinemas" in the 80s, should be seen not as a process of achieving "maturity" but rather as a series of transformations in which technology and its users attempt to achieve balance. This concept better explains the uniqueness of video diffusion in Poland, which did not follow a simpler, free market path. Difficult access to cassettes and VCRs meant that using video was not necessarily tantamount to owning a VCR. Thus, two conceptualisations of the technology and two corresponding phases of diffusion took place: first video was established as a collective and public medium, and then as a home medium.

Piotr Gaweł's approach to the development of the medium of video was based on political premises. He believed that the medium had a liberating potential, predominantly in terms of economics, restoring normal relations of supply and demand in the face of the hopelessness and backwardness of state production and distribution of goods.[49]

---

[49] See Piotr Gaweł, "Rynek video w Polsce. Próba analizy," *Zeszyty Prasoznawcze* 26, no. 3 (1985).

Its political potential is equally self-evident. As Gaweł concludes:

> The technical nature of video itself results in an individualised character of reception of the information conveyed through the VCR. Therefore, consumption is exceptionally decentralised. Furthermore, demanding a monopoly in the videocassette distribution, while being not fully justified legally, does not seem right from the point of view of general social benefits. A decentralised circulation is an objective necessity in the VCR development in every country, due to market essential flexibility and its ability to meet individualised consumer demand.[50]

Although the author adds later that creating private, legal video cassette rental shops would make them easier to control, it feels like an alibi for a clearly political programme postulating a free circulation of content and equating decentralisation with individualism. Such an interpretation of history, where video was a medium splitting the political and economic system of the People's Republic of Poland because of its "technical nature itself," passed into common perception.

Previously I have tried to question the teleological theories of media development, where the new naturally replaces the old. The problem comes back, however, in regard to the first video users. It has to be remembered that, for a long time, past even the mid-80s, they were not the same as owners. A model of private, individual ownership of a VCR used in the family circle or alone forms slowly and is anything but obvious.

The first video users did not own the devices. In the 70s thousands of video recorders were produced in Poland, and most of them were not acquired by private individuals, but rather by companies, cultural centers, and universities. The Łódź Film School television studio was equipped with VCRs. Video recorders were owned by many sport clubs (to record competitions and training sessions), police stations (along with cameras, they were used for professional purposes) and various other institutions. Apart from official use, they were also used in other ways. Private screenings were organised at work and the VCRs were even taken home for screenings organised in private apartments. Such an event is mentioned by the previously referenced Agnieszka Nieracka, whose husband used to bring home his university's VCR.[51] The VCRs were borrowed in a similar way – for private use – by the Łódź Film School employees.[52] Such practices were not unusual and many people came into contact with video in this manner before they had a chance to buy their own VCR.

---

[50] Ibid., 64.
[51] See Agnieszka Nieracka, interview, December 16, 2016.
[52] See Małgorzata Staszewska, interview, January 16, 2015.

Of course, VCRs were also bought by individual customers. Piotr Gaweł gave a meticulous typology of VCR buyers in the mid-80s. He divided them by age and financial resources and identified the first VCR owners as representatives of the oldest and most affluent group, whose savings exceeded $5,000.[53] He also noted that

> The average wage, equal (on the black market) to US$30–50 a month, did not allow for the accumulation of means that could be used to buy a VCR and excluded not only people with an average, but also a very high income from the group of potential buyers.[54]

Gaweł's perspective as a member of that period's society reveals at least some of the mechanisms of wealth inequality. In 1986 the cheapest VCR cost $400, roughly ten times the average monthly wage, but in spite of this the new technology was becoming popular. Who could afford a VCR?

Contrary to official declarations, the social system of the Polish People's Republic was characterised by substantial material inequalities and a paternalistic, quasi-feudal structure.[55] At the same time, those inequalities were partially masked by limitations of consumption. The consumerism of the Edward Gierek era had an egalitarian character, leaving the elites with considerably smaller possibilities for both conspicuous consumption and hoarding. Difficulties in travelling and buying real estate, currency, gold, and other traditional ways of accumulating wealth shifted the attention to luxury consumer goods that were available, at astronomical prices, in the shops of Pewex and Baltona, as well as on the black market. Electronic appliances, including VCRs, were found in this group.

Conversations with early VCR users clearly indicate that many of them were very affluent, often private entrepreneurs or members of privileged groups. Asked about the first VCR owners, Maciej Karwas replies:

> Private businessmen. Tailors, for example. I bought that Video-2000 tape recorder from a football player – it was Smolarek, a player from Łódź, who was friends with a group of tailors producing trousers, skirts, blouses and so on [...] Those private businessmen had a great need for viewing material: films, concerts, music, everything. It had to be organised somehow.[56]

---

[53] Gaweł, "Rynek wideo," 58–59.

[54] Ibid., 56.

[55] See Ireneusz Krzemiński, "System społeczny epoki gierkowskiej," in *Społeczeństwo polskie czasu kryzysu*, ed. Stefan Nowak (Warszawa: Wydział Filozofii i Socjologii Uniwersytetu Warszawskiego, 2004).

[56] Maciej Karwas, interview, May 15, 2014. In communist Poland private enterprise was limited to marginal areas of economy, like small-scale clothes manufacturing.

On the other hand Jacek Samojłowicz, later a businessman active in the video field and the founder of the Neptun Video Center distribution company, as well as a screenwriter and film producer, recalls the purchase of his first VCR as follows:

> I remember the first VHS recorders available in Poland; they were in the Pewex shops, were made by JVC, and I remember they were the VHS type, loaded from the top, and they cost $960. I remember I was sailing on ships at the time and in dollars I was earning $16–20 [a month] I think, which was typical for the period. But when [the VCR] came out, I had it the next day.[57]

Comparison of the price of a VCR and Samojłowicz's wages is surprising, but his profession as a sailor provided him with opportunity to make a lot of extra money, not entirely legally. Both interlocutors are among the privileged: Karwas because of his social and family relations, Samojłowicz because of his profession as a sailor. At the same time their situation differed from that of the tailors, as mentioned by Karwas. They both were somehow connected with film: Karwas professionally, Samojłowicz through a moviegoer's passion. This placed them in the second group of early VCR owners, namely enthusiasts, for whom the VCR was – as for the Workshop of the Film Form member, Wojciech Bruszewski – a tool of artistic expression, a means to realise a passion for cinema or a tool of professional activity. This second group of video users included cultural managers and student activists, who – like the previously mentioned Jacek Rodek – organised early screenings in student clubs, cinema clubs, and sci-fi clubs.

Finally, the third group of VCR owners were those connected directly with the communist regime. The anecdote cited by Stefan Szlachtycz, who was the main director of Telewizja Polska (Polish Television) in the years 1974–85, is characteristic in this regard. Szlachtycz recounts in an interview how Maciej Szczepański, head of the Radiokomitet at that time, built a distribution network of films and VCRs, to cater to the members of the Central Committee of the Polska Zjednoczona Partia Robotnicza (Polish United Workers' Party) and the party's regional secretaries. Telewizja Polska would acquire films from foreign distributors, copy them, then deliver to the dignitaries and their families once a week.[58]

The existence of Telewizja Polska's video library is well-documented but I have not been able to confirm the rest of this story. It is bizarre and at the same

---

[57] Jacek Samojłowicz, interview, October 10, 2014.

[58] See Stefan Szlachtycz, "Bufety na Woronicza," interview by Jacek Szczerba, *Gazeta Wyborcza*, accessed March 14, 2019, http://wyborcza.pl/duzyformat/1,127291,7717056,Bufety_na_Woronicza.html

time – what makes its authenticity even more dubious – fits perfectly into the Byzantine splendour of the regime in the Gierek era. However, if there is some truth to this tale, it would signify a very early adoption of video technology based on VCR, or even Video 2000 tapes. There is no doubt that people in the highest echelons of the communist regime had easier access to video through VCRs owned directly by the party. On the other hand, this advantage was diminishing with the diffusion of video technology.

Initially, video technology offered elitist consumption and was a means of hoarding. The former continued throughout the 80s due to economic instability and hyperinflation. There were, however, additional factors which contributed to the diffusion of video. The first and most important was the poor quality of the cinema and TV offer in the 80s. The economic crisis which had begun in the late 70s had a dramatic impact on film imports. Only 33 films were imported from the West in 1980, and in the first half of 1981, only two.[59] This was accompanied by a decline in domestic TV and film production, caused partially by financial difficulties, and partially by political pressures. Cinema attendance dropped dramatically. Therefore, the state stopped supplying its citizens with not only bread, but also circuses. And so, it is hardly surprising that the citizens took matters into their own hands.

This relative lack of entertainment was accompanied by another factor: the desire for technical novelty combined with consumer ambitions and the need for luxury, which was, essentially, an imitation of previous elitist consumption. Video became more a symbol than a definite set of technological possibilities, and the wish for entertainment was absorbed into the longing for political freedom and unrestrained consumerism. The VCR itself became one of the first objects of postmodern consumption, which takes place more in the symbolic rather than the real sphere.

Distribution development: sources of films and the emergence of Polish-language cassette editions

The diffusion of video technology in Poland was accompanied by an explicit belief that the country was lagging behind. This opinion was more or less explicitly expressed in the Polish press and even the discussion in *Film* magazine in 1970 was permeated by the idea that progress was taking place elsewhere; Poland, at most, could only observe the situation and try to prepare for changes.

---

[59] Edward Zajiček, *Poza ekranem. Polska kinematografia w latach 1896–2005* (Warszawa: Stowarzyszenie Filmowców Polskich i Studio Filmowe Montevideo, 2009), 271.

This is hardly surprising, because the country had little to say in the technological sphere, and subsequent imported innovations proved that the only effective
strategy was imitation supported by legal licensing or illegal industrial espionage.
At the same time, however, this official common-sense pessimism is surprising
(especially in retrospect) because it contradicts the ideological slogans about
the progress of the socialist economy. Apparently, this foundation of the political
system was extremely fragile.

The Polish media industry had to develop adaptation strategies, competing in areas where technological sophistication was less important and could be
compensated for in other ways. Unable to compete with innovations in the areas
of exhibition (data carriers and playback devices) or distribution (content sales
and rental system), the focus was on production – trying to provide content that
could be successful abroad. This explains the promotion, often effective, of Polish
filmmaking – especially art films. In the field of television production, this strategy also led to some success, as demonstrated by the increase in short film exports:
from 156 in 1961 to 821 in 1967.[60] The next stage of this strategy were the international successes of Polish children's television series (*Bolek and Lolek, Colargol,
Reksio* and others). The release of TeD discs with Bolek and Lolek films was an
early attempt to apply this strategy to video technology. Characteristically, these
small successes were all in the sphere of Marxist "superstructure" while the real
source of power, the industrial "base" (production and exhibition technologies)
was in the hands of the West. Pragmatism led to giving over control of an area
in which, according to Marxist theory, real changes and progress could take place.

In the quoted discussion from *Film* magazine, Andrzej Skawina, the then director of the state-owned film export enterprise Film Polski, saw video technology as another opportunity to implement the niche adjustment strategy:

> When asked whether the cassette cinema is our ally or enemy, I answer that
> it is a great ally for film export. It creates new possibilities for selling films abroad.
> How to sell? Everyone is still waiting. The current system of selling licenses for five
> years becomes obsolete because it is impossible to control. Anyway, new problems
> are emerging. For example, a film sold for cassette distribution cannot be sold to
> the same country's television; however, the reverse can happen.
>   We have already been asked for an offer. For now, we are not thinking about
> the production of films specifically for cassettes. The suggestions we receive for now
> are for already-made films.
>   The most concrete proposals are made by the Japanese. They want to take our
> films and bear all the costs associated with their release to the Japanese "cassette

---

[60] See Edward Zajíček, "Rozwój organizacyjno-gospodarczy kinematografii," in *Historia filmu polskiego*, vol. 5, ed. Rafał Marszałek (Warszawa: Wydawnictwa Artystyczne
i Filmowe, 1985), 304.

market." They would divide the profits between themselves and us. The company in the Federal Republic of Germany is interested in entertainment programmes with standardised duration. The cassette will cover a one-hour programme: 55 minutes of entertainment and five minutes of cassette manufacturer's advertising. There is no time limit in the United States. They are interested in feature programmes; they want to preserve all old colour films.

Personally, I see much greater sales opportunities for cassettes not with entertainment films, but with what we call the "quality films" offer. So far, the situation looks like this: to introduce a film to a country, one needs a minimum of potential viewers. This minimum is quite high – in some countries, several hundred thousand viewers. With distributing films on a cassette, this barrier decreases. The volume of cassettes, just like the volume of records or books, will be regulated depending on the market needs. Reaching, for example, the US market is extremely perilous. Cassettes can change a lot.

This opens up huge opportunities for exporting, especially among the Polish communities abroad, and without huge marketing expenses.[61]

Unfortunately, these hopes and aspirations remained unfulfilled. Film Polski managed to sell a small number of television programmes to Telefunken and Teldec but the few releases of Polish films on cassette in the 1980s were not a success among either the Polish communities abroad or among a wide range of foreign recipients. When the video offensive was launched by large Hollywood studios, their productions dominated the market, leaving space for, at most, exploitation films from smaller distributors – but certainly not for the offer that Polish cinema could have presented.[62] Instead of becoming a cassette films exporter, Poland experienced a demand which the domestic film industry was not able to satisfy.

When the state companies were either not willing or not able to explore the market possibilities of the new medium of video, private entrepreneurs took over. Initially, the demand for video material was small, limited to a tiny group of VCR owners, but even at this early stage specific properties of video technology became important. Video was radically different from, on the one hand, other

---

[61] "Film w kasecie – kino w domu" (discussion), *Film*, July 26, 1970: 6–7.

[62] This idea was constantly coming back. In 1986 a video club was launched at the Polish Cultural Centre in Sofia, Bulgaria, but it suffered from a lack of cassettes. (See ZAR, "Polskie video w Bułgarii," *Ekran*, January 21, 1988: 4). A few years later, when after the fall of communism private initiatives came to the fore, Krzysztof Rondo, the founder of the Video-Rondo company, spoke in an interview in 1990 about extensive plans to export VHS tapes with Polish films to Germany and to the East, to Lithuania and Russia (See "Krzysztof Rondo, Rodzinny interes. Rozmowa z Krzysztofem Rondo," *Cinema Press Video*, no. 8 (1990): 6).

hoarding methods used by the few financial elites of the Polish People's Republic, and, on the other, from luxury consumer goods such as food, alcohol, or even music records and music cassettes. The use of the video recorder as a home source of non-televised entertainment required, firstly, access to audiovisual materials, primarily feature films, and secondly, the Polish-language version of these materials. This issue often appears in the memories of the innovators. For example, Jacek Samojłowicz comments on it:

> Because I am from the coastal area, I knew a gangster whose nickname was Nikoś, who had a brother in Sweden, and he brought a whole video rental shop here. I was doing Polish versions at that time, because I knew the language. I translated them with a friend from high school, so I can say, more or less, what year it was, around the mid-70s. I remember the music at that time, Sweet, Slade, the others. My father was a sailor, so when these first video recorders showed up, we bought films in Great Britain. We translated them with a friend who knew the language even better than me and he did the voice-over, or we'd record them together. The people who bought them were mainly priests and wealthy people. Because there were no rental shops or anything like that, it was only this recording and voice-overs that became... a sort of market of Polish versions. So the first one... it could be called the beginning of a rental shop, but it was not a rental shop, it was only adding a Polish lector. For example, you brought your film and this film would be translated for you or someone already had the translation to sell you... But there were also official companies, I remember, there was one on Abrahama Street in Gdynia, and somewhere also in Warsaw. The practice was semi-illegal and people ended up being prosecuted, but that was not until the 80s.[63]

This statement certainly contains some chronological inaccuracies. Nikoś's brother, who brought cassettes from Sweden, probably did not do so in the mid-70s, because at that time in Poland the demand for such articles was still very low, and in Sweden there were no cassette rental shops; it must have happened at least a decade later. However, the anchoring of the processes described by the interlocutor during high school is rather indisputable and indeed falls within the first half of the 70s (Jacek Samojłowicz was born in 1956). Samojłowicz's account perfectly captures the essence of the cassette distribution process, namely the necessity to create Polish film versions, which at first was just a fun activity for teenagers (translating and exchanging films between friends), and gradually became an economic activity. The key element was access to films (importing them from Great Britain, with the help of the father who was a cargo ship sailor), language skills, and technical facilities in the form of at least two video recorders, thanks to which a lector's voice could be added to the soundtrack. Another important

---

[63] Jacek Samojłowicz, interview, October 10, 2014.

factor was the business acumen of the interlocutor, who earlier in the conversation mentions how he combined his passion for music with earning money by selling vinyl records imported from abroad. This is probably why he dates the events with reference to bands. The spontaneous remark from Samojło-wicz that his activity was "semi-illegal" is also accurate. It reflects the extra-systemic nature of these operations.

Another one of the video innovators already quoted in this book, Maciej Karwas, also describes this early period of video diffusion. He situates it at the end of the 70s and identifies the first owners of video recorders as private entrepreneurs:

> There was a huge demand for films from private entrepreneurs, who were keen on movies, concerts, and music; pretty much everything. It had to be organised somehow. I was lucky that my sister had been living in England, in London, since 1968. Because we kept in regular contact then, by mail, of course, or by phone, at the moment when it became possible, I asked if she could start recording TV programmes on the VCR and send them here. The only problem was with getting them to me, because they could not be sent through the official channels. Just like now, such parcels went to the customs office, and they checked what was recorded on them. Since it was recorded in a language other than Polish, it was immediately suspicious and deleted. [laughs] It was obvious that the content was foreign and hostile. So we made an arrangement with sailors who took the ferry between... between England and Gdynia. The sailors brought over these tapes. It wasn't just me who used their services, but they all... Jacek Samojłowicz, for example, who is a film producer today, ran the Neptun Video Center... My sister recorded individual cassettes, and when they got here, they were translated. It couldn't be done earlier... Besides, British television, with the exception of BBC1 and BBC2, had programmes interrupted by commercials, so the materials had to be edited to remove them.[64]

Karwas' account is similar to Samojłowicz's and typical for this period of video history in Poland: access to a narrow network of recipients; personal contacts allowing a person to obtain materials intended for further copying; knowledge and skills necessary to prepare Polish versions of cassettes and to copy these cassettes. Karwas mentions not only films: "and concerts, and music; pretty much everything." This suggests that, at least at the beginning, the concept of video as cassette television was also taken into account.

Both interviewees cite the 70s as the beginning of their activities. The demand for videos grew rapidly at the beginning of the next decade. Karwas recalls: "In 1980 it was, kind of, the peak with these films, because Solidarity had started."[65]

---

[64] Maciej Karwas, interview, May 15, 2014.
[65] Ibid.

Political factors were clearly important, copying tapes for the underground publisher Videonowa was part of Karwas's activity as a distributor in the 1980s. More importantly, the Solidarity movement brought about freedom in many areas, particularly in culture, and the circulation of video cassettes increased.

As demand grew, hobbyists became professionals. Karwas mentions importing recorded tapes not only with the help of his sister, but also through other regular, albeit illegal channels, from the UK and Germany. With the growing number of films, their localisation also improved. Karwas, himself a sound film director, quickly mastered the technical aspects of VHS and his wife, Alicja Karwas, a professional dialogist working at the Studio Opracowań Filmów (Film Post-Production Studio) in Łódź, helped him with other issues. Karwas acquired better equipment: stereo VHS video recorders and professional Blaupunkt devices, and started employing translators and lectors. Localisation involved translating the dialogue and adding a lector's voice to a film's soundtrack, which was not difficult in VHS technology. Lectoring was, and still is, standard practice in Polish television so the audience was accustomed to it.[66]

The localised films then went to the market, either directly to the shows organised by the distributors (more on these later), or through distribution networks. Maciej Karwas describes it as follows: "I had friends who came to me and ordered ten, fifteen cassettes, and then they distributed them further. In any case, I was not standing in the bazaar... and I delivered them in selected quantities."[67]

Jacek Samojłowicz recalls:

People came in great numbers. After all, there were no cell phones at the time. As for the more affluent ones, we went to them... So few people had the recorded films that we somehow all seemed to know each other... And it lasted until... when these places became... they transformed into sort of pseudo-rentals, recording shops. We used to call it 'recording of the Polish version,' meaning that anybody could bring a cassette for us to voice-over.[68]

Samojłowicz, like Maciej Karwas, provided cassettes for further distribution, and arranged places where it was possible to "record the Polish version," which were in fact an early form of rental shops where one could exchange blank cassettes for films.

---

[66] While Polish television have always used lectoring, films in cinema exhibition have been usually subtitled. VHS films were closer to television than to cinema in this respect.

[67] Maciej Karwas, interview, May 15, 2014.

[68] Jacek Samojłowicz, interview, October 10, 2014.

Karwas lived in Łódź and Samojłowicz in Gdańsk but establishments where Polish versions were made and films were distributed operated around the country, especially in other large cities. Robert J. Szmidt reports that in the famous student club Pałacyk in Wrocław, run by the Polish Students' Association, there was a studio in which cassettes were localised.[69] The importance of student culture in organising video screenings is well documented, and I will explore it further later. However, Szmidt's account points to another aspect: the collecting of films, preparing the Polish versions and exchanging them. The management of the Pałacyk asked Szmidt to build them a collection:

> They gathered some funds and they said: build a video library for us. So I started to buy cassettes and expand it, getting them from Pewex, usually. Later, because I sold that video recorder, I bought a second one in order to be able to copy films from one cassette to another at home. At that moment I became a man who had two video recorders, which was awesome. There were maybe three such people in Wrocław. But at first, I didn't trade at all… until we started to organise video exchanges. Video exchanges were happening around 1984, just when the atmosphere relaxed a little. Then the influx began, this first wave of players and so on.[70]

The mechanism described by Szmidt is slightly different than in the case of Karwas or Samojłowicz, who were focused on commercial distribution. Szmidt was primarily involved in the organisation of video screenings. Compiling the video library and preparing Polish-language versions of films was supposed to enrich his offer but was not necessary, especially because as he recalls, many films were accompanied by a voice-over that was read live by a lector during the screening.

Szmidt's account combines systemic and extra-systemic elements of different degrees of legality. On the one hand, an official cultural institution decided to allocate funds for the creation of a video library. On the other, an individual with a private collection was necessary, who thanks to his own entrepreneurship, collected films, exchanging and importing them from abroad, and at the same time acquired blank cassettes in Pewex, paying for them with illegally obtained hard currency.

In addition to import, a major source of films on video cassettes was Telewizja Polska (Polish Television): not the programmes it broadcast but its video film library created by Maciej Szczepański, who was the head of Radiokomitet between 1972 and 1980. Even if Stefan Szlachtycz exaggerated his description of the Party film distribution network, there is no doubt that television film resources

---

[69] Robert J. Szmidt, interview, April 25, 2015.
[70] Ibid.

leaked outside the walls of the station headquarters. Jacek Rodek, who worked in the student club Hybrydy in the 70s, recalls:

In 1979, I met a guy from Unitra and bought a video recorder for a VCR system from him. After that, I met another guy in Hybrydy, from TV, one of the production managers. He had access to the so-called "Szczepański's Film Archive." It was one and a half thousand films recorded on the VCR system. Most were junk, generally speaking, hopeless, nothing special. But there was… Because at that time I was interested in cinema, I knew what was good and what wasn't. And from this one and a half thousand, I chose about fifteen to twenty films in total… I still can list these titles, more or less, because they were all films that had been shelved [stopped by the censor], and what's more, they were very good films. There were titles like *The Tin Drum*, *The Deer Hunter*, *The 120 Days of Sodom* by Pasolini, Borowczyk's *Immoral Tales*, *Salon Kitty* – this was a film about a brothel in Berlin… for Abwehr, where the Gestapo planted a bug, a romantic story, but an interesting film, I would say. There was a film from Vietnam about… a motorcycle gang that came back from Vietnam, and the action takes place in the US. I don't remember the title at the moment, because the film was typically commercial. And from science-fiction there was, above all, the famous *Dr. Strangelove, or: How I Learned to Stop Worrying and Love the Bomb*, and I also got one porn flick that was hugely popular, namely *Deep Throat* with Linda Lovelace. And there was also… I don't remember the titles at the moment, but… Ah! *Emmanuelle* was there, the first one, of course… I think I also had *A Clockwork Orange*, Fellini's *Satyricon*, but I'm not sure.[71]

This quote lists many of the big hits of early video distribution in Poland. It is no coincidence but rather the result of Jacek Rodek's undertakings during this period. Although he did not create a distribution network at that time, and did not prepare Polish-language versions of films, he organised video shows all over the country during which the cassettes were copied for further distribution. Rodek was not alone. The cassettes from the "Szczepański Archive" likely leaked through other channels and the same title could have several Polish-language versions made in different places.

Jacek Samojłowicz mentions yet another source of films, namely recording them from satellite television:

When this breakthrough came about, and VHS appeared, we knew that we needed content. Well, this content… First, the content was… I don't remember what was first. I remember satellite TV. The first satellite channel was Filmnet, and in Poland in the communist era you had to have an official permit to own an antenna, so there may have been only a hundred of them. My dad was one of the owners, and I bought one too. To get better reception, I gave the antenna to a friend who was

---

[71] Jacek Rodek, interview, February 4, 2015.

a translator, to install it in Szczecin, because the quality of the converters was not good, so the closer to the West, the better the picture was. And I remember that we were recording these films from Filmnet and translating them, and we were just... exchanging them with each other.[72]

This account certainly applies to the later period, because Filmnet began broadcasting in 1985 (first in Scandinavia). Apparently, even in the second half of the 80s, films were obtained this way. Samojłowicz, who did not produce Polish-language versions himself, stresses that his market advantage was access to new titles. He imported new films from abroad or recorded them from satellite television, ordering their Polish-language versions from companies in Warsaw, and then distributed the cassettes in the Gdańsk area.

Although video cassettes could be brought to the country by anyone in a more-or-less legal manner, their distribution was in fact based on a relatively small number of sources. Initially the cassettes came either from the leaks from the Polish Television video library ("Szczepański's Archive") or from private import channels, organised by a few distributors as a development of their hobbies in the late 70s.

In 1980 Maciej Szczepański was removed from his position in Radiokomitet and his TV video archive lost its significance due to new video formats. After the declaration of martial law in December 1981 the borders were closed and importing films became extremely difficult. Polish ports remained the only sources of black market trade. Even after 1983, when the borders were partially reopened, foreign travel was limited. The most popular and accessible black market routes connected Poland with other communist countries but attractive films could only be imported from Western Europe (later also from Asia).[73]

The second barrier to the mass import of video cassettes was the language barrier. Translation (usually by ear), editing and condensing the dialogues, professional lectoring and sound recording – all this required a complex network of social connections and professional facilities. In the meantime, competition had developed quite rapidly in the video market, favouring those distributors who imported a lot of films and were able to produce high-quality Polish-language versions.

Maciej Karwas, who is quite critical about his beginnings in the video business, also gives an example of his poorly-prepared competitors:

There was a guy who recorded somewhere near Wąchock. He became famous for lectoring porn flicks using the so-called onomatopoeic sounds, which means that when a German woman shouted 'Aaah! Aaah! Tiefer, tiefer!', meaning 'Deeper,

---

[72] Jacek Samojłowicz, interview, October 10, 2014.
[73] Jacek Rodek mentions the import of films from Singapore.

deeper!', he thoroughly substituted everything, including all the grunts. The absolute peak of his activity, so to speak, was dubbing all the vocal parts in *Fiddler on the Roof*. He sang all the vocal parts [laughs], voicing the actors, Topol and other known… well-known Western actors. He really put his soul into it but it came out terribly awkward, unfortunately.[74]

There is a widespread opinion that video had a democratising effect in the communist countries, that it shattered old cultural hierarchies and celebrated popular tastes. Karwas's anecdote suggests that even if the content was new, old hierarchies remained, at least on a technical level. Plebeian taste could be revealed shamelessly but the language version had to be executed professionally.

The work of translators is important in this respect. One of them, a Łódź University lecturer working for Maciej Karwas, Janusz Wróblewski, recalls:

There were three basic procedures, which went like this: sometimes Maciek Karwas showed me a film on video at his place, then he gave me an audio cassette only, because it was simpler to work with. For one thing, I did not have a VCR at home yet at that time; for another, it's hard to stop and rewind a video cassette. I mean, the tape always moved for a few additional seconds. So, I watched the film on video, then I had the soundtrack on audio, and I worked from that soundtrack. I stopped the tape and I produced at once a translation from what I heard. The second version was when they gave me only an audio cassette and wanted me to translate without having seen the film. We did this when Maciek didn't have time. After two films I categorically refused because it didn't make any sense. That was impossible, and besides… I remember two specific cases […] And sometimes, maybe somewhat later, I was given transcripts, except that I remember that some of the transcripts were also not fully accurate. I guess the transcriber didn't understand the dialogue very well.[75]

At the time, Wróblewski already had a lot of experience as a translator because he previously translated films screened at the Łódź Film School, where he met Karwas. Interestingly, the translator usually watched a film only once and then worked without a video player. This, of course, may explain some of the mistakes in the translations, although when the number of translators and translated films increased, a lack of competence was probably the main reason for errors. The quoted fragment, above all, indicates a period in the history of video technology when it is not yet a home medium, and even those directly involved in the development of local versions of films did not have easy access to it.[76]

---

[74] Maciej Karwas, interview, May 15, 2014.

[75] Janusz Wróblewski, interview, June 7, 2014.

[76] Translation of film dialogues from audio cassettes is also mentioned by another translator, Piotr W. Cholewa, who lives in Silesia (Piotr W. Cholewa, interview June 18, 2014).

Wróblewski also complains about renumeration – it depended on the run-time of the film, not on the amount of text or the difficulty of the translation. He was aggrieved by this: as a highly skilled linguist he usually received films difficult to translate and was not adequately rewarded. Wróblewski, however, worked in quite unique, professional conditions, facilitated by Maciej Karwas. His translations were, for example, edited by expert dialogists. Other studios often employed amateurs or spent less time polishing the dialogues.

Cooperation with professional translators was not always effective, as demonstrated by one of Karwas's anecdotes. He employed the eminent translator and writer Zbigniew Batko, whom he commissioned to prepare a translation of a Monty Python film:

> Zbyszek Batko translated *Monty Python and the Holy Grail* brilliantly, except that it was completely unusable. I told Zbyszek that if we are to translate the script, he must compress this translation to fit in the subtitles. The subtitles cover roughly a third of the text spoken by the actors, so you need to choose carefully, conveying what is essential for the viewer to understand. With lectoring, you can do more, but it is also limited to the fact that the lector should read only when the actor speaks on the screen. Zbyszek said that this is impossible because they have such wonderful dialogue and clever puns, and he must translate it all. To this day I have this translation, even though it could not be recorded or converted into subtitles due to its length. It's a shame, because aside from being unusable, it's a sensational translation.[77]

Cassette cinema

In the 1980s in Poland video was not individualised, home-owned entertainment, but rather a medium of public screenings. This situation was not at all obvious. A few decades earlier, television also went through a similar phase, but it was considerably shorter, and individual use was much more common from the beginning.[78] In Western Europe and in the USA video became a home medium almost instantly and the social networks of videophiles were built upon the exchange of tapes and not on sharing of equipment.[79] In Poland it took a decade for VCRs to become standard home appliances. The most important reasons were the price and availability of video players. When the format war began and the VCR system was going out of use, Polish manufacturers could

---

[77] Maciej Karwas, interview, May 15, 2014.

[78] Public screenings of satellite television are described in the last part of this book.

[79] See Joshua M. Greenberg, *From Betamax to Blockbuster. Video Stores and the Invention of Movies on Video* (Cambridge, MA: The MIT Press, 2008).

neither keep up with technological changes nor afford new licenses. Unlike with TV sets, Poles were left with very expensive imported VCRs, available for only the wealthiest. The comparison with television is striking. Production of TV sets started in 1956 and in the first decade their number increased from zero to 2.5 million.[80] In 1983, ten years after the first MTV 10 Polish video recorder was produced, there were between 20,000 and 70,000 VCRs in the country. Their number reached a million only in the late 80s. A quick look at these numbers may suggest that video diffusion was rather slow and that the television revolution, organised by the state, proved to be much more effective than the bottom-up, deconcentrated introduction of video recorders. This conclusion would be further strengthened by a comparison with the US market, where in 1989 67.6% of households with TVs also had a video recorder, compared to 1980 when this percentage was only 2.4%.[81]

The real picture is, in fact, more complex and the comparisons of video and television, as well as between Polish and American markets, have their limitations. Market saturation with TV sets was growing when state television provided more broadcasts and while the price of a TV set was admittedly high, it was within the financial reach of at least the wealthier sections of society.[82]

In contrast to this, in the 1980s domestic production of video recorders almost stopped, state-owned film companies did not supply films on cassettes, and attempts to create a distribution network through state-owned rental shops proved to be a misfire. The state did not help the video industry grow, but, in fact, suppressed it by chaotic legal and fiscal action. Unlike with television, the communist regime saw no propaganda opportunity in video. Under these circumstances, the development of the video market was quite significant.

Comparison of Polish and American markets, on the other hand, emphasises different diffusion patterns in these countries. In the USA VCR manufacturers had promoted their devices for home use from the very beginning. This resulted in the conceptualisation of the medium as a home cinema.

This model of video use could not be widely adopted in Poland because VCRs were too expensive. Users had to construct another concept of the medium to reconcile their needs with their economic possibilities, while taking into account technological conditions. Under these circumstances the concept of video as "cassette cinema" emerged. This model existed for the whole decade of the

---

[80] In 1966, there were exactly 2,540,100 television subscribers in Poland. See GUS, *Mały rocznik statystyczny 1967* (Warszawa: GUS, 1967), 209.

[81] See Frederick Wasser, *Veni, Vidi, Video. The Hollywood Empire and the VCR* (Austin: University of Texas Press, 2001), 68.

[82] In 1966 a TV set cost three and a half times the average annual salary in Poland. Twenty years later the cheapest video recorder cost ten times the average salary.

1980s, through public screenings organised in various places, from private homes to regular cinema halls, either free or ticketed. The shows allowed large audiences to gain access to films on cassettes, but also to get acquainted with the technology itself. The few VCRs around were put to good use and the number of cassettes was minimised. Localisation of films was also easier, since often, especially in the first half of the decade, foreign-language versions were screened with lectors reading the scripts live, or even translating on the spot.

This conceptualisation of video should not be seen as a "transitional" or substitute mode, just as early cinema cannot be treated as a "transitional" phenomenon. Admittedly, cassette cinema did not last long but magnetic tape video technology itself was not particularly long-lived. Video recorders went out of use at the beginning of the twenty-first century, so the period of cassette cinema covers at least a quarter of their entire history in Poland. Rather than presenting this history as a deterministic path towards the global model of home use, a local specificity should be included.

Public video shows had varied in scale: they could attract from a few to a dozen to over a thousand participants. Regardless of the size of the audience, its participants often complained about poor conditions: bad quality of the copies, small screen, muffled sound, faulty translations, and finally, stuffy rooms with uncomfortable, hard chairs.

For example, Jolanta Filar, who studied in Warsaw in the early 80s, recalls the screenings that took place in a student dorm:

> At the turn of the 70s and 80s we were watching films on TV sets. Or rather, on a TV set! It was not two people, but two hundred people watching a film on a TV set... For example, in the dormitory Hermes there was a hall, a large common room. Well, not really large, there were no giant rooms in the dorms. But there was a TV set... What TV? I don't remember, no more than 21 inches for sure. I don't know how large the screens were back then. Maybe it was a Russian one, I don't remember, one of the larger, colour ones. The TV stood on a table and people sat on the floor, then on chairs, and the last row sat on the tables. It was quite far away; I remember that I couldn't get a closer seat, to watch *Caligula*. Of course, the sound was loud so we could hear, and there was a lot of people. I remember, the temperature was probably thirty-two degrees, even though outside it was cold; it was winter then. But we were simply choking in there. I still remember the quality of it on the TV. It was actually more listening than watching, but everyone usually knew what was going on; knew more or less what the content was and who the actors were. Even if someone sat far away, they knew what they were watching. It didn't bother us.[83]

---

[83] Jolanta Filar, interview, October 31, 2014. Maciej Karwas made a Polish-language version of *Caligula* using a cassette imported from Germany. Perhaps, the copy Jolanta Filar saw in Hermes came from him.

The low quality of the image surprised not only the viewers, but even the organisers themselves. Robert J. Szmidt recalls:

Later, when we, as the Student Council, organised such screenings, we once went to Jelcz, near Wrocław, to a military garrison, because they wanted to screen *Conan the Barbarian* for the soldiers... I brought the equipment and we set up a black and white TV set, because it was impossible to connect a colour one there. Besides, it was a SECAM and the films were in PAL, so it had to be black and white. So I turned it all on, they let the soldiers in and I left because I didn't want to stay there. But I had this gut feeling, so I came back in to the other end of the room, which was a decent sized canteen. I stood at the back, among these soldiers in the last row. Well, I am sorry to tell you, the picture was like this [he puts two fingers together]. Knowing this film, I knew what was supposed to be on the screen, but the people who stood there were looking at the equivalent of a moving postage stamp.[84]

Although these screenings created a sense of community among viewers, their atmosphere was rather serious – they were not social gatherings for which the film would be just an excuse. Ryszard Borys, who participated in screenings organised by the Silesian Science Fiction Club in the 80s recalls:

It was not like watching a football match together, where we would talk freely and drink beer. No, no, no. We were focused on watching the films. Even the bad films were, at least until the sound was good, given our full attention. They were seen as works of art.[85]

Apparently, despite poor conditions, a special etiquette was observed, not unlike cinema theatre etiquette.

Why were these screenings so attractive? What made more than a hundred students crowd into a narrow common room, if they could spend time much more comfortably at the cinema, not to mention any other entertainment? Why did the viewers, who were accustomed to perfect screening conditions with large cinema screens and clear sound, gaze at the blurred image and listen to the screeching of a Russian 21-inch TV set?

To answer these questions, one should recognise the advantages in the limitations of video screenings. Poor picture quality meant that video was not an illusionist medium at the time.[86] The low resolution of small screens, faded copies,

---

[84] Robert J. Szmidt, interview, April 25, 2015.

[85] Ryszard Borys, interview, June 18, 2014.

[86] It can be argued that video, like traditional analogue television, has never been an illusionist medium, and the low technical quality of the image has always given it,

and image synchronisation problems meant that viewers severely, but perhaps also ecstatically, experienced the material presence of the medium.

When new technologies compete with old ones, they are not necessarily better in every respect. They have both advantages and shortcomings. Each innovation is progressive in some aspects, but regressive in others, and they both depend on the value system within which they are defined. For some, the innovation itself is attractive. Innovators and early followers are ready to endure many inconveniences just to experience something that is fresh and new. This explains the satisfaction of early video screenings, which offered collective access to the new medium. The technology itself was more important than the films on the screen, so it had to be experienced somewhat painfully.

Typically, the interviewees often remembered the technical aspects, such as the format of the cassettes, their shape, the size of the TV screens, and all the functions of the video recorders. This was just as important as the films themselves and applied not only to those who organised the screenings but also ordinary viewers. Of course, this effect wore off over time, which is why the recipients gradually attached more importance to the quality of the screening conditions and the artistic or entertainment value of the films. Interest in the novelty gradually diminished.

The technological novelty was obviously not the only advantage of the early video screenings. Participation in extra-systemic practices and the sense of freedom they gave were also important. This was especially the case of the screenings of overtly political films. An example of this was the first Videonowa film released on cassette, *Przesłuchanie* (*Interrogation*), directed by Ryszard Bugajski. *Przesłuchanie* was finished in December 1981 but it was banned and not officially released until 1989. In 1982 Bugajski managed to make a U-matic copy of his film and Videonowa released it on VHS. *Przesłuchanie* quickly became a hit at underground video screenings and an important part of the political experience of many viewers during this period.[87] Soon, it was joined by other independent productions, particularly those of Video Studio Gdańsk, among others.

Even if a film was not overtly political, the mere fact of participating in a public screening, of watching with others a film that was banned (or at least not approved) by the authorities, was community-forming. The lack of physical

---

referring to McLuhan's terminology, the coolness that demanded warming up by the recipient's imagination.

[87] See Ryszard Bugajski, *Jak powstało Przesłuchanie* (Warszawa: Świat Książki, 2010), 99. Bugajski gives a detailed account of the film's creation and its illegal distribution. He also quotes a characteristic anecdote about a viewer who was surprised to find out that *Przesłuchanie* was shot in colour, because after the video screening he was convinced that it was a black and white film.

or perceptual comfort viewers might have experienced was rewarded with a sense of independence and belonging. Unlike the enchantment of innovation, this experience did not wear off over time. For some people, it was an extension of their politics or religion, while for others, it was simply a hobby (as in the case of science fiction fan clubs). Most often, however, this sense of being "outside of the system" was associated with consumer desires. As a result, consumerism appeared as an extra-systemic attitude, and excited viewers could watch, for instance, the adventures of Chuck Norris or erotic films, while experiencing a sense of existence outside of the communist system, or even of a noble fight against it.

This does not mean, however, that the films themselves were irrelevant. If that were the case, distributors would not have tried to import new titles. On the one hand, the video offer was shaped by political circumstances, as shown by the example of *Przesłuchanie*, and by other popular anti-communist films at the time, such as *The Deer Hunter* (Michael Cimino, 1978). On the other hand, it was formed by the tastes of various segments of the audience. They were often completely different, but because distributors tried to satisfy them all, their offer often contained a mixture even more diverse than the one chosen by Jacek Rodek at the headquarters of Telewizja Polska, from erotic films, through to cinema classics, to art films and religious films. As the market developed, distributors expanded their offer and viewers became pickier. The poor quality of entertainment offered through official channels, in cinemas and on television, made the task easier, especially in the first half of the 80s.

The cassette cinema model was not modified. As the charm of the novelty diminished, the importance of the technical quality of the screenings increased. The once-important sense of the extra-systemic nature of such activities also decreased. Eventually, though not immediately, the matter was resolved by the new Film Act, adopted on July 16, 1987. The new law included video, finally resolving the question of videocassettes' copyright. The Act regulated both public screenings and the sale and rental of cassettes. A loophole was left for free screenings and free exchange of films, which helped cassette cinemas survive for some time in parishes where the screenings were not ticketed, and in places such as science fiction fan clubs, where screenings were club activities and were financed through membership fees.

The new Act led to a short-term formalisation of the cassette cinema model. Commercially operating "video clubs" could be created provided that they gained permission from the state Film Committee. "Video cinemas" operated on similar terms. For example, in 1988 such a cinema was opened in the Katowice Centre of Soviet Science and Culture. The cinema was equipped with two halls, and it had a rental shop providing cassettes free of charge to the members of the Polish-Soviet Friendship Association. The offer, of course, included Soviet films and was not a roaring success.[88] Some regular cinemas started adding video screenings to their offer. It was obvious, however, that the cassette cinema model was a thing of the past, soon to be replaced by a home video model.

Video exhibition circuits

The first public video screenings were organised by cassette distributors. They often started their business activity by organising screenings, and only then, based on accumulated cassettes, became distributors. This was the case of Jacek Rodek, who acquired his first package of films on cassettes to screen at the Hybrydy student club. When the shows proved to be a success, Rodek launched them in other cities:

> These films, or the set I had, was very popular because clubs from various cities invited me. In Wrocław, we did shows for 3,000 people for three consecutive days. There were one, probably two a day, where in three days around 10,000 people came and watched.[89]

The screenings took place at the Pałacyk student club mentioned earlier. Rodek brought his video player there, but the club quickly acquired its own device:

> In Wrocław, at the Pałacyk, I brought my video player to the first screening, and they arranged the rest of the equipment: the Rubin monitors, because back then the only colour TV set was Rubin. They borrowed them from various sources: from the Party Committee, from the community centre, and from other places. But for the next screening, which took place three or four months later, I didn't have to bring the VCR anymore, because they earned so much from the first one that they already had their own equipment. I only brought the films.[90]

I described the rest of this story above: Pałacyk not only bought its own video recorder, but when the VCR format went out of use, the club employed Robert J. Szmidt, who built a distribution and exhibition network there.

The material and organisational basis of these earliest public screenings was provided by a network of student clubs cooperating with each other under the aegis of the Polish Students' Association. The clubs could officially combine business (shows were ticketed) with the statutory cultural activities. They also

---

[88] See Irena Białek, "W Katowicach nowe kino," *Ekran*, January 21, 1988: 30–31.
[89] Jacek Rodek, interview, February 4, 2015.
[90] Ibid.

provided relative protection against possible sanctions related to the illegal organisation of cinema screenings, which was explicitly prohibited by the Film Act
from 1951.[91] Perhaps the existence of these legal provisions helped in the cultural formation of the new medium of video, because the organisers of the screenings were eager to emphasise how much it differed from cinema. Jacek Rodek
also mentions inviting the authorities to these shows, to protect the organisers
against possible sanctions: "Of course, there were not only regular people at these
screenings, students and so on, but there were also VIP rooms for people from
censorship, from the Party Committee, or simply for VIPs."[92]

Screenings in student clubs were hugely popular, as Jacek Rodek recalls:

> There were so many invitations at one point that I couldn't manage to go to them
> all. I would have to travel all over Poland, because they invited me everywhere to
> do such screenings, so I shared the work. I started to pass it on to other colleagues
> from Remont, for example, and also from the electroacoustic division, because we
> knew each other.[93]

Gradually, the video screenings went beyond the student club network.
Examples include the previously described show organised by Robert J. Szmidt
in a military garrison in Jelcz, where he screened *Conan the Barbarian*. Maciej
Karwas, in turn, was asked to organise a screening of James Bond films at the military garrison in Solina.

In the 1980s, regular video exhibition circuits required a combination
of three factors: organisational support from some existing institution, access to
rooms where the screenings could take place, and a reliable, permanent audience.
Apart from the country-wide network of student clubs, two other video exhibition circuits were most important: science fiction fan clubs and Catholic parishes
(and other Church institutions).

For obvious reasons, there was no permanent, separate exhibition circuit
through which overtly political films could be systematically presented to a specific audience. Cassettes published by underground video publishers circulat-
ed around the country, but they were usually watched at home, at ad hoc video
screenings, or on Church premises. They also contributed, sometimes heavily, to
the offer of the other circuits.

---

[91] Article 14. of the Film Act threatened that anyone could be arrested for up to
a year or pay a fine for operating light theaters or public screenings of films. However,
the authorities were reluctant to apply this law from the Stalinist era, so although its letter
was unambiguous, no one was convicted of breaking it.

[92] Jacek Rodek, interview, February 4, 2015.

[93] Ibid.

Science fiction fan clubs

Science-fiction literature was published in the Polish People's Republic from the 1950s, but it boomed two decades later. In the 1970s science fiction fandom emerged. In 1976 Ogólnopolski Klub Miłośników Fantastyki i Science Fiction (National Fantasy and Science Fiction Fan Club) was founded, formally as a branch of the Polish Students' Association. Four years later the Klub was replaced by the Polskie Stowarzyszenie Miłośników Fantastyki (Polish Association of Science Fiction Fans).[94] The Association connected several branches and independent clubs in the country, facilitating domestic, and even foreign, contacts.

As a result, a network of science fiction fan clubs was created, with thousands of fans forming a potential video audience. The structure of the clubs also facilitated the establishment of the cassette cinema model: they had a room in which regular meetings were held, and they also had access to video recorders, sometimes borrowed from various institutions, sometimes private. Moreover, some of these clubs had already held film screenings. For example, Piotr Rak, a member of the Silesian Science Fiction Club, remembers Russian films borrowed from the embassy, which were screened using an 8 mm projector. It must have been an important part of the club's activity, since the club had bought a projector and two club members, including Rak himself, took a course to learn to operate it.[95] Video screenings were seen as a modern development of these 8 mm projections. They also benefited from a similar privilege, being regarded as part of club activities rather than just commercial film shows.

The common interests of SF club members facilitated the establishment of video distribution connections. It was also important that, organisationally, the SF fan movement originated from the same source as the other video exhibition network, i.e. from the student club movement. The key figure in this respect was Jacek Rodek, who combined interests in both video and science fiction, and had been active in both areas from the beginning. As an employee of the Hybrydy student club, Rodek organised two large festivals of science-fiction films in the early 80s, which were important, perhaps even formative, events for SF fandom in Poland.[96] Video copies of fantasy and science fiction films began cir-

---

[94] The Polish term "fantastyka" refers to science fiction, fantasy and other speculative fiction genres.

[95] Piotr Rak, interview, June 18, 2014. Although the interlocutor clearly speaks about the 8 mm projector, it is possible that, in fact, it was a 16 mm projector – that is what another interlocutor, Piotr W. Cholewa mentions.

[96] Robert J. Szmidt mentions in an interview that he came to one of these festivals from Wrocław, spending the night at a train station, and watching films screened in Hybrydy during the day.

culating around the country, commonly copied at fan conventions or exchanged between clubs.

Piotr W. Cholewa, an eminent translator and fan club member, describes the role video played at fan conventions:

> At conventions, a cassette copying studio was, for a long time, one of the key elements. For many years, until 1988, maybe even 1989, such film rooms were crucial; that is, these were the most crowded events. Typically, there were about ten video recorders and you could copy what they had there or what someone had brought.[97]

Over time, video screenings that had initially only been an attraction, became the most extensive part of the official programme. Piotr W. Cholewa recalls that the convention "Dni Fantastyki" in Dzierżoniów in 1983 caused a sensation among fans because it announced the screening of films from video cassettes. And, it was Jacek Rodek's films that were to become the main element of the early repertoire of fantasy fan clubs: *Blade Runner* (Ridley Scott, 1982) and *Star Wars – Episode VI: Return of the Jedi* (Richard Marquand, 1983).[98] Interestingly, the latter film was officially released on VHS in 1986, so the copy watched at the convention must have come from the television video library.

Science fiction film festivals were also organised, such as "Lapsbiax" in Gdańsk (in 1985 and 1986) and "Przegląd Filmów Fantastyki" ("Review of SF and Fantasy Films") in Świnoujście (from 1984) or "Od Mélièsa do Spielberga" ("From Méliès to Spielberg") in 1983 in Łódź, but they usually presented 35 mm copies.

Video screenings were typically held in clubs scattered throughout Poland as part of their regular activities. For example, Śląski Klub Fantastyki (Silesian Science Fiction Club), one of the most important in the country, started video screenings in 1984. Initially they were organised ad hoc, but quickly began to occupy a permanent spot in the club calendar, taking one evening a week, and even two evenings a week for some time between 1985–1986, once the club's film committee was formed.[99] The first film shown on video at the Silesian Science Fiction Club was *Star Wars – Episode VI: Return of the Jedi* – probably a copy of the cassette watched a year earlier in Dzierżoniów.[100] The film was provided

---

[97] Piotr W. Cholewa, interview, June 18, 2014.

[98] Ibid.

[99] The club chronicle printed in the magazine *Fikcje*, published by the Śląski Klub Fantastyki, informs that from October 1985 there would be two screenings on Fridays: the main at 6 pm, and the second, at 8 pm, with films which go beyond the categories of science fiction or fantasy.

[100] The magazine *Fikcje* states: "7-10 12 – *Return of the Jedi, The Thing, Conan* and *Blade Runner* at video screenings" ("Kronika klubowa," *Fikcje*, no. 11 [1984]: 2).

by a distributor from Warsaw, working with Jacek Rodek. Jerzy Ferencowicz, a member of the club, recalls:

> It was the last part of *Star Wars*, which arrived without a Polish-language version, and the two people who took this copy came, and tried to record the lector's voice on this unprofessional equipment. It was, so to speak, almost impossible.[101]

Poor equipment, the indisposition of the lector, and a poor location stood in the way: the club was, at that time, located next to a railway line. Ultimately, however, the lectoring worked out, and the screenings were a great success: "We had a full house for three weeks... and word got out in Katowice, that you could watch video films in the Śląski Klub Fantastyki."[102] Eventually, the club gave up on making Polish-language versions of films and relied on distributors; sometimes original-language video films were screened with simultaneous translations. This indicates a general tendency: science fiction fan clubs did not develop their own distribution except for ad hoc copying of films during conventions. Instead, they created a strong exhibition network focused on science fiction and fantasy films.

Regular screenings at the Śląski Klub Fantastyki were an important element of its activity. The significance of the screenings is demonstrated by the fact that the club set up a special screening room, equipped with about 50 old cinema chairs. At first, video recorders were brought together with the films, then they were borrowed from members, and eventually the club bought its own device. The Helios TV set was the club's property. As a rule, separate screenings for club members were followed by discussions, and open, ticketed screenings were held for a wider audience. The latter screenings were preceded by a lecture on the film. On the one hand, it was a form of promotion for science fiction, and the club itself. On the other, they followed the model of the Film Discussion Club meetings, which were still popular at the time (and always included a lecture). There were also closed screenings of films less related to science fiction or fantasy, but with a clear political profile, such as *Red Dawn* (John Milius, 1984).[103] However, my interlocutors from the club unanimously emphasised that openly political topics were to be avoided. The club's activities were controlled by the secret agents of the Ministry of Public Security, of which everyone was quite aware, so

---

[101] Jerzy Ferencowicz, interview, June 18, 2014. Ferencowicz was an important figure in this period of the club's activity, with a high membership number: 006.

[102] Ibid. Piotr W. Cholewa remembers, however, that the first screenings in the club were run by a private company hired by the Śląski Klub Fantastyki. This company also brought its own equipment.

[103] See Ryszard Borys, interview, June 18, 2014.

they tried not to cross the dangerous line. Nevertheless, video screenings took place in the grey area, which was relatively free.

The films reached the club through video distributors, usually Robert J. Szmidt, who was a member of the Śląski Klub Fantastyki, as well as Jacek Rodek, who often came from Warsaw. Cassettes were also copied and exchanged during conventions, and some were even imported from abroad through private contacts and the international links of the SF fandom.[104] The club exchanged materials with the Czech fandom, so films from Czechoslovakia were screened. Examples include *Adéla ještě nevečeřela* (*Dinner for Adele*, Oldrich Lipský, 1978), *Srdečný pozdrav z zeměkoule* (Oldrich Lipský, 1982), and several others.

Over time, the Śląski Klub Fantastyki began to compile its own video film library, focused on science fiction and fantasy films. The library operated under loose rules. As club members recall, cassettes were not borrowed to take home, but sometimes one could copy them on the spot if one brought one's own video recorder. Later, according to information from the *Fikcje* magazine, members were also allowed to exchange cassettes. This indicates a rather long period of the extra-systemic status of video; even when the cassette cinema model was already relatively stable, cassette circulation remained unregulated. The club film library was also used for specific, quasi-distribution activities, namely organising video screenings in other institutions, such as schools or cultural centres. Combined with the lectures, film screenings contributed to the club's budget:

> In addition to these, say, fees, we organised events: for example, an event for miners, for some community centre and so on. We were paid for this, which means that we officially got money for delivering a lecture, because earning money from film screenings wasn't allowed. The films we had on video helped us a lot, because as we went to talk to the head of the cultural centre or whatever and we said, "we also have some films," and we'd reel off the titles. Of course they were curious, which is how we knew we'd got the job.[105]

However, this activity did not turn into any regular form of distribution, and the film library itself, which already had several hundred films at that time, was unfortunately plundered by thieves during looting of the club's premises in the late 80s (both the video equipment and cassettes were valuable to the thieves). During the next decade, the collection was partially restored, and the video film library operated as a rental shop for members of the Śląski Klub Fantastyki.

---

[104] Ryszard Borys reports that many cassettes for the club were bought during a trip to the Worldcon (The World Science Fiction Convention) in Brighton in 1987. Piotr W. Cholewa, in turn, brought a copy of the film *Dune* (David Lynch, 1984) from England.

[105] Andrzej Kowalski, interview, June 18, 2014.

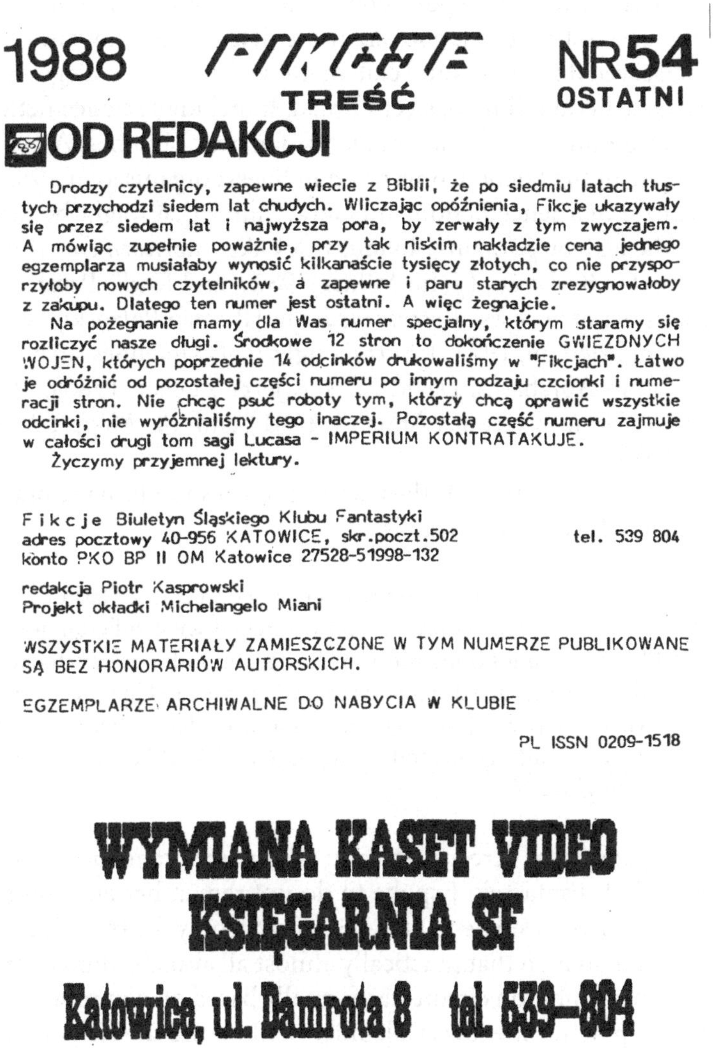

1988 *FIKCJE* NR**54**

TREŚĆ OSTATNI

OD REDAKCJI

Drodzy czytelnicy, zapewne wiecie z Biblii, że po siedmiu latach tłustych przychodzi siedem lat chudych. Wliczając opóźnienia, Fikcje ukazywały się przez siedem lat i najwyższa pora, by zerwały z tym zwyczajem. A mówiąc zupełnie poważnie, przy tak niskim nakładzie cena jednego egzemplarza musiałaby wynosić kilkanaście tysięcy złotych, co nie przysporzyłoby nowych czytelników, a zapewne i paru starych zrezygnowałoby z zakupu. Dlatego ten numer jest ostatni. A więc żegnajcie.

Na pożegnanie mamy dla Was numer specjalny, którym staramy się rozliczyć nasze długi. Środkowe 12 stron to dokończenie GWIEZDNYCH WOJEN, których poprzednie 14 odcinków drukowaliśmy w "Fikcjach". Łatwo je odróżnić od pozostałej części numeru po innym rodzaju czcionki i numeracji stron. Nie chcąc psuć roboty tym, którzy chcą oprawić wszystkie odcinki, nie wyróżnialiśmy tego inaczej. Pozostałą część numeru zajmuje w całości drugi tom sagi Lucasa - IMPERIUM KONTRATAKUJE.

Życzymy przyjemnej lektury.

F i k c j e Biuletyn Śląskiego Klubu Fantastyki
adres pocztowy 40-956 KATOWICE, skr.poczt.502   tel. 539 804
konto PKO BP II OM Katowice 27528-51998-132

redakcja Piotr Kasprowski
Projekt okładki Michelangelo Miani

WSZYSTKIE MATERIAŁY ZAMIESZCZONE W TYM NUMERZE PUBLIKOWANE SĄ BEZ HONORARIÓW AUTORSKICH.

EGZEMPLARZE ARCHIWALNE DO NABYCIA W KLUBIE

PL ISSN 0209-1518

WYMIANA KASET VIDEO
KSIĘGARNIA SF
Katowice, ul. Damrota 8   tel. 539—804

Fig. 4. Advertisement for the exchange of video cassettes in the Silesian Science Fiction Club

Source: *Fikcje*, no. 54 (1988): cover

Club video screenings featured both older films, such as *Dr. Strangelove, or: How I Learned to Stop Worrying and Love the Bomb* (Stanley Kubrick, 1964), and newer ones, from the final part of the *Star Wars* trilogy, to the extremely popular *Blade Runner* and the adventures of Indiana Jones. The full list of films is difficult to reconstruct, since regular club screenings were hardly ever documented.

At most, a monthly fanzine report, concise and cursory, came out. Convention programmes and festival catalogues are more detailed, but not always reliable.[106] Both the preserved documents and club members' memories suggest that the category of science fiction films was very broad. It included science fiction, fantasy, horror, but also films with a wide range of "fantastic" themes. A good example is the catalogue of the Silcon Convention in Silesia organised in 1985, which includes a lexicon of science fiction and fantasy films that might be interesting to the fans.[107] The lexicon contains obvious titles like *Capricorn One* (Peter Hyams, 1978), *Conan the Barbarian* (John Milius, 1982), and *Krull* (Peter Yates, 1983), but also a number of horrors (*Rosemary's Baby*, [Roman Polański, 1968], *The Exorcist*, [William Friedkin, 1973]). Surprisingly, it also includes such diverse films as *Airplane!* (Jim Abrahams, David Zucker, Jerry Zucker, 1980), *Quest for Fire* (*La Guerre du feu*, Jean-Jacques Annaud, 1981) and *The Muppet Movie* (James Frawley, 1979).

This rather broad notion of "the fantastic" corresponds to the practice of film screenings. Jerzy Ferencowicz recalls:

> It wouldn't be true if I said that in the science fiction club – Śląski Klub Fantastyki, which deals with science fiction – we only watched science-fiction movies. No. We watched films with James Bond too. One Friday a month was devoted to martial arts films, and those were martial arts films that were made in Hong Kong, with German dubbing, where a guy was jumping out... not Jackie Chan, because it wasn't Jackie Chan, but he was jumping out and shouting: *Hände hoch! Halt!* So the whole room roared with laughter.[108]

The programme of screenings came about spontaneously. The members of the Śląski Klub Fantastyki I spoke to do not remember any controversial issues. There was also no single person responsible only for the selection of films. The demand was so high that practically almost all available films were watched. So, although the exhibition circuit was formally based on common interest in science fiction, in practice the low availability of films and the lack of thematic specialisation of distributors meant that a very wide range of films was offered, from science fiction and fantasy films, to action films, political films and erotic films.[109]

---

[106] Jerzy Ferencowicz mentions cases where a particular film could not be obtained, although its title was already printed on the tickets and in the programme.

[107] Mariusz Ramach, Tomasz Szczepański, "Mały leksykon filmów fantastycznych," in *Silcon85. Informator konwentowy* (n.p., 1985).

[108] Jerzy Ferencowicz, interview, April 25, 2015.

[109] Robert J. Szmidt and Elżbieta Gepfert both mention the viewing of erotic and pornographic films at conventions. See Robert J. Szmidt, interview, April 25, 2015, and Elżbieta Gepfert, interview, June 18, 2014.

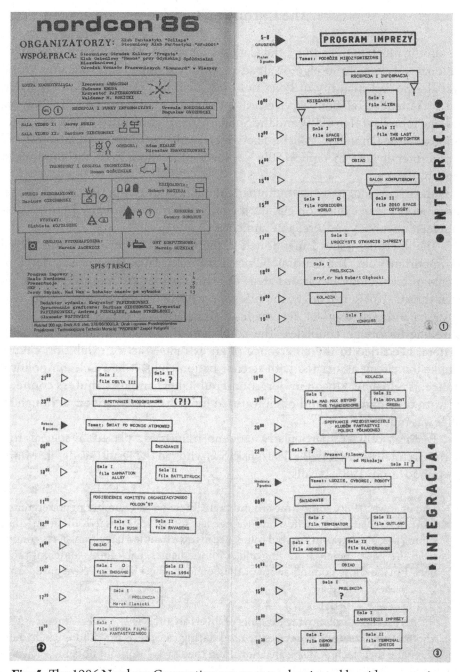

Fig. 5. The 1986 Nordcon Convention programme dominated by video screenings

Source: Krzysztof Papierkowski, ed., *Informator I Konwentu Klubów S-F Polski Północnej Nordcon'86* (n.p., 1986), 1–3

The Catholic Church

In the nineteenth century, the Catholic Church became a symbol of Polish national resistance to foreign occupation and it continued this role after World War II. The vast majority of Poles not only identified themselves as Catholics, but also attended religious services. After the election of Cardinal Karol Wojtyła as Pope John Paul II in 1979, and the Solidarity revolution in 1980, the role of religion and the Church in Polish society increased. The Catholic Church offered – both literally and metaphorically – a space of freedom from the communist regime.

The network of parishes and other church institutions perfectly met the requirements for an exhibition circuit: it combined organisational support, access to screening rooms, and a reliable audience.

The doctrinal issues I mentioned in the first part of this book also played an important role. The Catholic Church has been favourable to the developments of technology in various areas of life, stressing the opportunities rather than threats. Not surprisingly, the priests, especially young ones, were often innovators and early users of new media.[110] They also had contacts abroad, often travelling to foreign countries. The combination of these circumstances with pastoral needs led to the emergence of an extremely active exhibition circuit. Unlike the circuit associated with science fiction clubs, it had a clearer political stance: it relied heavily on underground publishers and distributors connected with the opposition, and often featured films that were banned, although its offer was much wider.

Pastoral centres, particularly student ministries, played a special role in the circuit. Father Jacek Pleskaczyński S.J., who ran the Jesuit Academic Ministry in Łódź from 1985, recalls:

> Maybe it was 1986, some time around then. There were a number of underground films in circulation, cassettes, and we needed something to watch them on. They came via a fairly straightforward route, because we handled large transports coming from the West, mainly from England, with medicines, baby foods, different products. I made those coming aware of the demand. There was an active group from

---

[110] An excellent and moving example of this is an interview with Fr. Zbigniew Mistak, who first enumerates the devices he used during catechesis for children and young people, giving their technical specifications, and then says: "Without noticing what is in this world, without referring to it, without building relations with this world, you cannot love God, the creator, can you? That's the way to put it, no? " and then he spontaneously recites the canticle "All Creation Worships God" (Fr. Zbigniew Mistak, interview, May 29, 2015).

England that I met later. I visited them there. They secured the importation of such equipment without any trouble: a video recorder and a large monitor by the standard of the time, so that it was possible to have screenings in a large room, which was my intention. I also asked for a camera, which later turned out to be indispensable because at a later date it was used to film the 1987 pilgrimage of the Pope, now Saint, John Paul II to Poland, when he was in Łódź.[111]

Fr. Pleskaczyński S.J. came to Łódź in 1985, so obtaining the video recorder, which the Academic Ministry did not have then, was one of his first undertakings. He used a charity channel to acquire the equipment, which proves that he made this acquisition a priority. His own experience was not without significance here, because in his school days he ran the Film Discussion Club, and later created one at the Academy of Catholic Theology. Therefore, he must have understood and appreciated the formative importance of cinema and audiovisual communication.

However, the use of video equipment was not only the domain of large pastoral centres in big cities. Another of my interlocutors, Fr. Zbigniew Mistak, became a parish priest in the village of Krzątka in the Podkarpacie region at around the same time. He had previously used traditional film technique (8 mm tapes) in his catechetical work, and he describes his parish's video equipment:

> The video recorder appeared when I got my own parish. At that time, I was a priest and a teacher, a catechist… I think some time around 1986–1987. Back then, catechesis was taught outside of school hours, so a priest who had appropriate premises was at an advantage. My first parsonage was a new church in Krzątka, a hall, where I created there a sort of catechetical workshop. Of course, there was a library with various songbooks, with texts from the Holy Bible, but I gave up these rather primitive materials that were worn out and falling apart. There was also a screen for films and then came a video recorder. A video tape player, and a video tape recorder that could record from the television… So at that time, I bought a TV set for this room… the largest screen that was available then, with a diagonal of 28 inches [laughs]. Back then, the TV sets were small, but for the room, I thought that the children must have a decent picture.[112]

Both priests were keen on having their own equipment and screening films as professionally as they could. They also rapidly incorporated new technology into regular catechetical and pastoral activities. The video was also a development of both priests' earlier interests and skills. The introduction of video took place

---

[111] Fr. Jacek Pleskaczyński S.J., interview, February 2, 2016.

[112] Fr. Zbigniew Mistak, interview, May 29, 2015. The entire conversation suggests that the room could have been equipped with video equipment even earlier, in 1985.

around the mid-80s, so slightly later than in the case of the Śląski Klub Fantastyki, but it is difficult to draw a more general conclusion about the belated development of the parishes' exhibition circuit.

In parishes and ministries, video equipment was used less systematically, and there were no regular screenings (at least in the places I describe here). In the Jesuit Academic Ministry, evenings during the week were filled with meetings and seminars, and screenings were usually organised on Sunday evenings, but not weekly. Fr. Mistak from Krzątka in Podkarpacie recalls it similarly – in addition to using video for catechetical purposes, he also organised screenings for parishioners in the church hall. However, there were no designated days or times. When he happened upon an interesting film, the priest invited the community.

The screenings were, of course, free of charge and were very popular, at least initially:

> We had a relatively large hall and sometimes, with some films, the place was packed to the gills. I had to repeat some of the screenings… As for the seats, there are maybe… I haven't been there in a long time… one hundred and several dozen, or maybe near two hundred, but it was generally necessary [to put the screen] somewhere high… on a stage, and even then, on some kind of a platform.[113]

Similarly popular were video screenings in Krzątka:

> I invited people from my parish to come watch films at my parsonage. It was probably the first such event for them. People came to the lower church, which needed to be adapted a little so people could sit on benches and see the TV. It was big, so we set it up on a table. Anyway, I know that people were excited about it. It was such a novelty, wasn't it? They watched films that no one else was showing.[114]

Both interlocutors emphasise the technical details of the shows, the spatial organisation of the rooms, and the adaptation of these rooms to frequent screenings. This confirms my thesis that public screenings were not a secondary experience, complementing the usual cinema theatres offer or acting as a substitute for home cinema. It was simply a stable, independent social concept of this technology, constituting a complete medium for the users.

Film screenings were accompanied by discussions or lectures. In the case of Fr. Pleskaczyński S.J. the idea of lectures was clearly borrowed from his previous experiences with the Film Discussion Club, so that in his accounts of these

---

[113] Fr. Jacek Pleskaczyński S.J., interview, February 2, 2016.
[114] Fr. Zbigniew Mistak, interview, May 29, 2015.

two areas overlap. Through discussions and talks, priests also tried to bring out the religious values of films, not limiting themselves to focusing on their entertainment or even artistic merits.

In the case of the parish exhibition circuit, it is much more difficult to indicate the origin of films. The irregularity of screenings was probably not conducive to constant contact with distributors, and as a result the cassettes came from various sources. However, the early video distributors certainly cooperated with the Church. Maciej Karwas mentions that he did work for the Łódź convent of Discalced Carmelites, from whom he received films, including archival titles "from the beginning of the century," some on 16 mm tape. Karwas copied these films to magnetic tape, edited them and returned a Polish-language version copy. Copying took place in the convent, from where the cassettes were further distributed by the Carmelites.[115]

Contacts within the political opposition and, more broadly, various social, artistic and other circles that operated outside the state-controlled sphere played a large role in the development of this exhibition circuit. The Church of the Holy Name of Jesus in Łódź, at which the Jesuit Academic Ministry was based, led the pastoral care of artists. The Jesuits also worked closely with opposition activists. These environments intermingled, which resulted in greater availability of various materials on video. Fr. Pleskaczyński S.J. mentions that the films were obtained through the same channels as books and magazines published underground, and in particular he recalls the cassettes issued by Videonowa. Video recordings of *Dzwonek Niedzielny* were also screened.[116] In addition to this, strictly political programmes and religious films dominated. Fr. Pleskaczyński S.J. recalls:

> People were mad for these films. I remember the one by Bugajski, with Janda… *Interrogation*, well, this type of film was screened again and again. Another such hit was *Fiddler on the Roof*. There were not many religious films; they were often of poor quality, but sometimes there were some. These were not very professionally

---

[115] See Maciej Karwas, interview, May 15, 2014. Interestingly, Karwas copied these materials in the Polish Television Łódź studio, using professional equipment. The materials went to the television archive and later were used by the station.

[116] Fr. Pleskaczyński not only screened *Dzwonek Niedzielny* but also participated in filming these video programmes: "It just so happened that when I had the video camera, sometimes I also filmed or cooperated in filming these programmes *Dzwonek Niedzielny*. Of course I was by no means a professional… But it so happened that sometimes I went to Warsaw. I obediently dealt with my workload in the church on Sunday [in Łódź], and then we got in a car and drove to Warsaw, because it was usually filmed, if I remember correctly, late in the afternoon or evening." See Fr. Jacek Pleskaczyński S.J., interview, February 2, 2016.

made productions, but *Jesus of Nazareth* by Zeffirelli was very well received, and it-was a double experience. After all, such religious films were not watched in Poland at that time, there was no place for them.[117]

Other films that were screened include *Pielgrzym* by Andrzej Trzos-Rastaw-iecki (*The Pilgrim*, 1979) and the recordings of subsequent pilgrimages of John Paul II to Poland. In turn, Fr. Zbigniew Mistak recalls the Hollywood classics: *Ben-Hur* (William Wyler, 1959), *The Robe* (Henry Koster, 1953), *The Ten Commandments* (Cecil B. DeMille, 1956) and *Quo Vadis* (Mervyn LeRoy, 1951), but also *One Flew Over the Cuckoo's Nest* (Miloš Forman, 1975) and films by Andrzej Wajda: *Człowiek z marmuru* (*Man of Marble*, 1977) and *Człowiek z żelaza* (*Man of Iron*, 1981). Fr. Pleskaczyński's remark about a small number of religious films sounds quite surprising. Considering the titles he mentions, like *Fiddler on the Roof* (Norman Jewison, 1971) and even *Rambo – First Blood* (Ted Kotcheff, 1982), it suggests that due to the low availability of cassettes and always getting them from the same distributors, various exhibition circuits featured the same titles, and as a result the parish circuit was open to a very wide spectrum of films.

Conclusion: the decline of cassette cinema and the birth of the video rental shops

The material factors supporting the existence of cassette cinema were the high price and scarceness of video recorders. Another factor was initially the lack and then the weakness of officially operating sources of cassettes, whether for purchase or to rent. The psychological factor was the sense of community built independently of, or contrary to, the official system of entertainment. By the end of the 80s all these factors gradually lost their significance. Video recorder prices started to fall, and buying them was easier as official and unofficial import increased. In 1989, a used video recorder could be purchased on the black market for PLN 2.5 million, while the price of a VHS cassette ranged from 30 to 40 thousand.[118]

The state rental network was struggling. In March 1988 there were 32 rental shops in the country, and the offer included only 90 Polish and 30 foreign films. The price of the cassettes themselves was also a barrier – because

---

[117] Ibid.

[118] See MAKLER, "Co. Gdzie. Za ile?," *TOP*, December 1, 1989: 4. In 1989, the average monthly salary was PLN 206,758 ("Przeciętne miesięczne wynagrodzenie w latach 1950–2008," accessed December 13, 2019, https://www.infor.pl/prawo/zarobki/zarobki-w-polsce/686166,Przecietne-miesieczne-wynagrawał-w-latach-19502008.html).

they were quite expensive, state-owned rentals had only 14,000 in total.[119] At the same time, however, private video rentals developed, and this market was finally regulated by the aforementioned Film Act of 16 July 1987. Before that, their legal status was unclear. Based on this law, as reported by the magazine *TOP*, the Film Committee received 200 applications for permission to run a rental and for the distribution of cassettes, and 80 such permits were issued.[120] Therefore, the offer of available films whose legal situation was fully regulated expanded. This concerned not only Polish but also foreign films, as private companies started to buy rights to distribute films on cassette. ITI was leading the way with its contract with Warner Bros.[121]

By the end of the 80s in Poland, according to various estimates, there were from several hundred thousand to one million video recorders.[122] The desire to have video at home became commonplace. In 1986, the *Ekran* magazine introduced the "Video-Ekran" [Video-Screen] section, initially still quite modest in volume and rather critical. It included, for example, an interview with Lesław Wojtasik entitled "Videodywersja" [video-sabotage]. Photographed in a shirt and holding a book, Wojtasik was presented as a professor, rather than a military general, which he was at the time. The quasi-civilian Lesław Wojtasik warned his readers about the information war waged by the US using video:

> We are observing the strategy deployed when anti-socialist "art" created in the past was first popularised. For example, *Doctor Zhivago*, filmed twenty years ago, very quickly became available on cassette. Now films made with the intent of political sabotage include Vietnam movies – *The Deer Hunters*[sic], some three episodes of *Rambo* and *Missing in action*. Other films released in this way include *Gorky Park*, some episodes from the James Bond adventure series, and so on.[123]

A year later, *Ekran* magazine was publishing completely different articles – descriptions of 17 more official video rental companies operating in the country, information on the prices of video recorders, as well as regular columns, such as "ABC Video" (describing the operation of video recorders), "Eksplozja video" ("Explosion of Video," discussing artistic and social effects of video), "Video-notki" ("Video Notes") and "Poczta A–Z video" ("A–Z Video Mail"). The latter

[119] See f, "Wypożyczalnie video," *Ekran*, March 23, 1988: 5.

[120] tl, "Video," *TOP*, March 4, 1988: 20.

[121] See "Ofertę Warner Bros przedstawia dyr. Andrew Somper," *Ekran*, October 29, 1987: 10–11.

[122] See Karol Jakubowicz, "Video: wróg czy przyjaciel telewizji," *Ekran*, April 14, 1988: 2.

[123] Lesław Wojtasik, "Videodywersja. Rozmowa z prof. dr hab. Lesławem Wojtasikiem," *Ekran*, August 3, 1986: 2.

column consisted of answers to readers' questions, usually related to technical issues, although other problems were also covered. For example, 14-year-old Adaś wrote that he was saving his pocket money for a video recorder and asked for advice on which device to buy. The editor put a dampener on the boy's enthusiasm, but it was a clear sign of rapidly growing consumer aspirations.[124]

The change in the video usage model, which is, in fact, the redefinition of the medium, was gradual. This was accompanied by various adaptation practices, such as renting video recorders or sets consisting of a video recorder and TV set, usually to watch together among friends.[125] Conversations with participants of such shows confirm that they were exceptional events – not in terms of frequency, but in terms of rituals involved. They were organised on Saturdays or Sundays, much less often in the middle of the week, and they were associated with social gatherings, which further reduced the cost per participant. Rental companies not only provided the equipment and cabling, but also cassettes, and delivered everything to people's homes.

Throughout the 80s, a transition between public and home screenings emerged. Piotr Gaweł, proposing a typology of public shows, gives two varieties of "absolutely closed" screenings: free and commercial.[126] Gaweł's typology is very debatable, because he obviously (and consciously) omits screenings of a political nature. The category of "closed shows" is nevertheless very instructive. These shows can be seen as a hybrid "home cassette cinema": using the home space, but opening it, to various degrees, to an outside audience. This is confirmed by a fascinating account courtesy of Janusz Wróblewski, the translator mentioned earlier:

> Once I was driven by car to, I think, Tomaszów Mazowiecki. [...] It was simply a private screening, if I remember correctly, for some, I don't know, people like party activists. I was supposed to produce a live translation of one of the *Emmanuelle* films. [...] I had not recorded it prior so I simply translated it live...[127]

This screening took place in the early 80s. It was not unique, but was a standard part of the activity of one of the distributors in Łódź, who delivered not only

---

[124] See "Wakacyjna poczta A–Z Video," *Ekran*, August 20, 1987: 31.

[125] In 1987, it cost PLN 2.5 thousand per day to rent a video recorder. The price of the video recorder at that time was about half a million zlotys, although they were also bought for dollars. See Lilla Lesiak, "Co? gdzie? kiedy? za ile?," *Ekran*, January 8, 1987: 31. The much lower price of the video recorder than given earlier for the year 1989 results from hyperinflation.

[126] See Gaweł, "Rynek wideo," 68.

[127] Janusz Wróblewski, interview, June 7, 2014.

cassettes to people's homes, but also a translator. Clearly, already then, cassette cinema merged with video home cinema.

After 1987, during the transitional period between the two models, the sources of cassettes did not change. Cassette copying shops continued to operate. Gradually, they transformed into legally operating rental shops, although they still offered copy and translation services. Copyrighted films were legally copied as well. For example, the Poltel agency, together with ITI Video Service, offered a set of children's films, which could be copied for a fee to a cassette one had brought with themselves or to a cassette bought (for hard currency) in a Baltona shop. In the latter case, the customer received a discount for copying. Of course, this type of activity was only a step away from creating a rental company.

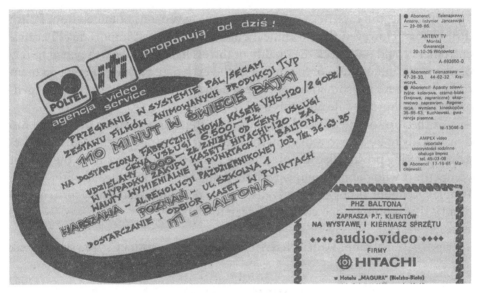

Fig. 6. Announcement about recording films onto cassettes

Source: *TOP*, February, 1987

As I mentioned earlier, video cassettes, even blank, were expensive due to the weakness of the Polish złoty. Setting up a rental company with an adequate number of titles was therefore a costly undertaking. Instead, cassette exchange points began to develop, where one could get a film by exchanging it for an old one. In this way, the owner avoided wearing out the cassettes and limited the number of them. Because these points operated locally and, to some extent, collectively (several entrepreneurs offering such services could be found in one place, typically a local market), each of them could offer a limited range of titles.

Maciej Przepiera, who at the end of the 80s run this type of business, recalls:

First, I was selling the players and videos that I brought over from Berlin. When I realised that I basically had free access to cheap equipment, I copied loads of cassettes and decided to do what everyone else was doing: rent them out. Back then, on these markets that operated in Gdańsk on different days just near where I live, there were a dozen or so such points and everyone did well… There was high demand. I mean, I must say that it wasn't the only source of video films, because then, as I recall, there were some official ones. They were more permanent in character, bricks and mortar, but they had less choice and higher prices, and they weren't run very well. But on the market it was different, people were only paying for the exchange. Still, to have something to exchange, they had to buy from me first.[128]

Of course, the cassettes were Polish-language versions obtained from Polish distributors. However, their technical quality was of great importance. Because customers could not check the quality of the exchanged videos, they had to rely on the word and opinion of the owner of the exchange business. Przepiera mentions that the militia sometimes intervened and requisitioned cassettes, but generally the authorities tolerated this business.

At the turn of the 80s and 90s, however, the video industry developed too dynamically to be able to remain in the grey area of extra-systemic activities. Legally operating private distributors (like ITI) started to remove from the market competitors who did not comply with the Film Act. In 1989 the fall of the Polish People's Republic brought the privatisation of reception, and a new form of medium – that is home video cinema – began to triumph.

Translation: Wojciech Szymański and Aleksandra Czyżewska-Felczak

Bibliography

Afanasjew, Jerzy. *Okno Zbyszka Cybulskiego*. Warszawa: Prószyński i S-ka, 2008.
Bachmann, Gideon. "Czy będziemy kupować filmy w sklepach." *Film*, May 2, 1965.
Białek, Irena. "W Katowicach nowe kino." *Ekran*, January 21, 1988.
Bugajski, Ryszard. *Jak powstało Przesłuchanie*. Warszawa: Świat Książki, 2010.
Cendrowicz, Paweł. "Fonica w świecie CD, czyli 'dyskofony' z Łodzi." Accessed November 7, 2018. http://www.technique.pl/mediawiki/index.php/Fonica_w_%C5%9Bwiecie_CD,_czyli_%E2%80%9Edyskofony%E2%80%9D_z_%C5%81odzi#Podsumowanie.2C_ocena_lub_rachunek_sumienia
Currie, Tony. *A Concise History of British Television, 1930 – 2000*, 2nd ed. Tiverton: Kelly Publications, 2004.

---

[128] Maciej Przepiera, interview, February 3, 2015.

"Film w kasecie – kino w domu" (discussion). *Film*, July 26, 1970.

Gaweł, Piotr. "Rynek video w Polsce. Próba analizy," *Zeszyty Prasoznawcze* 26, no. 3 (1985).

Gaweł, Piotr. "Rynek wideo w Polsce." *Film na Świecie*, no. 334–335 (1986).

Gaweł, Piotr. "Zasięg video w Polsce." *Zeszyty Prasoznawcze* 25, no. 3 (1984).

Greenberg, Joshua M. *From Betamax to Blockbuster. Video Stores and the Invention of Movies on Video.* Cambridge, MA: The MIT Press, 2008.

GUS. *Mały rocznik statystyczny 1967.* Warszawa: GUS, 1967.

Infor. "Przeciętne miesięczne wynagrodzenie w latach 1950–2008." Accessed December 13, 2019. https://www.infor.pl/prawo/zarobki/zarobki-w-polsce/686166,Przecietne-miesieczne-wynagrawał-w-latach-19502008.htmlf

Jakubowicz, Karol. "Video: wróg czy przyjaciel telewizji." *Ekran*, April 14, 1988.

"Kronika klubowa." *Fikcje*, no. 11 (1984).

Krzemiński, Ireneusz. "System społeczny epoki gierkowskiej." In *Społeczeństwo polskie czasu kryzysu*, edited by Stefan Nowak. Warszawa: Wydział Filozofii i Socjologii Uniwersytetu Warszawskiego, 2004.

Kuźmicz, Marika, and Łukasz Ronduda, eds. *Warsztat Formy Filmowej / Workshop of Film Form.* Warszawa: Arton Foundation / Fundacja Arton; Berlin: Sternberg Press, 2017.

Lesiak, Lilla. "Co? gdzie? kiedy? za ile?" *Ekran*, January 8, 1987.

MAKLER. "Co. Gdzie. Za ile?" *TOP*, December 1, 1989.

Mallinson, John C. "The Ampex Quadruplex Recorders." In *Magnetic Recording. The Frist 100 Years*, edited by Eric D. Daniel, C. Denis Mee, Mark H. Clark. New York: IEEE Press, 1999.

Neurobot, "Unitra." Accessed June 27, 2018. http://neurobot.art.pl/03/n-files/unitra/unitra.html

"Ofertę Warner Bros przedstawia dyr. Andrew Somper." *Ekran*, October 29, 1987.

Papierkowski, Krzysztof, ed. *Informator I Konwentu Klubów S-F Polski Północnej Nordcon'86.* N.p., 1986.

Ramach, Mariusz, and Tomasz Szczepański. "Mały leksykon filmów fantastycznych." In *Silcon85. Informator konwentowy.* N.p., 1985.

Robakowski, Józef. "Video Art – Szansa podejścia rzeczywistości." *Gazeta Szkolna PWS-FTViT*, 1976. Accessed January 2, 2017. http://repozytorium.fundacjaarton.pl/index.php?action=view/object&objid=3173&colid=75&catid=18&lang=pl

Rodek, Jacek. "Seks, kłamstwa i kasety wideo." In *Hybrydy. Zawsze piękni, zawsze dwudziestoletni*, edited by Sławomir Rogowski. Warszawa: Fundacja Universitatis Varsoviensis, 2013.

Rondo, Krzysztof. "Rodzinny interes. Rozmowa z Krzysztofem Rondo." *Cinema Press Video*, no. 8 (1990).

RTV. Radio i Telewizja, April 8, 1979.

Stempel, Wiesław. "Film w kasecie – co nowego? Mówi dyrektor Wiesław Stempel." Interview by Elżbieta Smoleń-Wasilewska. *Film*, November 22, 1970.

Stempel, Wiesław. "Pierwsze słowo techniki." *Film*, August 9, 1970.

Sugaya, Hiroshi. "Consumer Video Recorders." In *Magnetic Recording. The Frist 100 Years*, edited by Eric D. Daniel, C. Denis Mee, Mark H. Clark. New York: IEEE Press, 1999.

Sugaya, Hiroshi. "Helican-Scan Recorders for Broadcasting." In *Magnetic Recording. The Frist 100 Years*, edited by Eric D. Daniel, C. Denis Mee, Mark H. Clark. New York: IEEE Press, 1999.

Szlachtycz, Stefan, "Bufety na Woronicza." Interview by Jacek Szczerba. *Gazeta Wyborcza*. Accessed March 14, 2019. http://wyborcza.pl/duzyformat/1,127291,7717056,Bufety_na_Woronicza.html

tl. "Video." *TOP*, March 4, 1988.

Terramedia. "TeD video disc." Accessed December 12, 2016. http://www.terramedia.co.uk/media/video/ted_video_disc.htm

TOP, February, 1987.

Tosi, Virgilio. "Kiedy umrze kino." *Film*, January 11, 1970.

Unitra-klub. "Magnetowidy." Accessed April 12, 2019. http://unitraklub.pl/magnetowidy

Unitra-klub. "ZRK MTV-10." Accessed April 12, 2019. http://unitraklub.pl/node/1365

Urbański, Bolesław. *Telewizja kasetowa*. Warszawa: Wydawnictwa Komunikacji i Łączności, 1972.

Wajdowicz, Roman. *Historia magnetycznego zapisu obrazów*. Wrocław: Ossolineum, 1972.

Wajdowicz, Roman. *Nowoczesne metody rejestracji obrazów*. Warszawa: Komitet do Spraw Radia i Telewizji, 1975.

"Wakacyjna poczta A–Z Video." *Ekran*, August 20, 1987.

Wasser, Frederick. *Veni, Vidi, Video. The Hollywood Empire and the VCR*. Austin: University of Texas Press, 2001.

Wielage, Marc, and Rod Woodcock. *The Rise and Fall of Beta*. Accessed 11 December 2016. http://www.betainfoguide.net/RiseandFall.htm

Wikipedia. "Videotape Format War." Accessed December 12, 2016. https://en.wikipedia.org/wiki/Videotape_format_war

Wojtasik, Lesław. "Videodywersja. Rozmowa z prof. dr hab. Lesławem Wojtasikiem." *Ekran*, August 3, 1986.

"Wypożyczalnie video." *Ekran*, March 23, 1988.

ZAR. "Polskie video w Bułgarii." *Ekran*, January 21, 1988.

Zajiček, Edward. *Poza ekranem. Polska kinematografia w latach 1896–2005*. Warszawa: Stowarzyszenie Filmowców Polskich i Studio Filmowe Montevideo, 2009.

Zajiček, Edward. "Rozwój organizacyjno-gospodarczy kinematografii." In *Historia filmu polskiego*, vol. 5, edited by Rafał Marszałek. Warszawa: Wydawnictwa Artystyczne i Filmowe, 1985.

Interviews

Ryszard Borys, interview by Krzysztof Jajko and Piotr Sitarski, Katowice, June 18, 2014.

Piotr W. Cholewa, interview by Krzysztof Jajko and Piotr Sitarski, Katowice, June 18, 2014.

Jerzy Ferencowicz, interview by Krzysztof Jajko and Piotr Sitarski, Katowice, April 15, 2015.

Jolanta Filar, interview by Michał Pabiś-Orzeszyna and Piotr Sitarski, Warszawa, October 31, 2014.

Elżbieta Gepfert, interview by Krzysztof Jajko and Piotr Sitarski, Katowice, June 18, 2014.

Maciej Karwas, interview by Piotr Sitarski, Łódź, May 15, 2014.

Andrzej Kowalski, interview by Krzysztof Jajko and Piotr Sitarski, Katowice, June 18, 2014.

Fr. Zbigniew Mistak, interview by Piotr Sitarski, Stany, May 29, 2015.

Agnieszka Nieracka, interview by Piotr Sitarski, Wrocław, December 16, 2016.

Fr. Jacek Pleskaczyński S.J., interview by Michał Pabiś-Orzeszyna and Piotr Sitarski, Warszawa, February 2, 2016.

Maciej Przepiera, interview by Karolina Burnagiel, Gdańsk, February 3, 2015.

Piotr Rak, interview by Krzysztof Jajko and Piotr Sitarski, Katowice, June 18, 2014.

Jacek Rodek, interview by Piotr Sitarski, Warszawa, February 4, 2015.

Jacek Samojłowicz, interview by Piotr Sitarski, Łódź, October 10, 2014.

Jacek Siciński, interview by Piotr Sitarski, Łódź, March 8, 2017.

Małgorzata Staszewska, interview by Katarzyna Woźniak, Łódź, January 16, 2015.

Robert J. Szmidt, interview by Piotr Sitarski, Poznań, April 25, 2015.

Jerzy Wojtas, interview by Piotr Sitarski, Łódź, December 20, 2016.

Janusz Wróblewski, interview by Piotr Sitarski, Łódź, June 7, 2014.

PART III

Maria B. Garda

MICROCOMPUTING REVOLUTION IN THE POLISH PEOPLE'S REPUBLIC IN THE 1980S

PART III

Section Two

MICROCOMPUTING REVOLUTION
IN THE POLISH PEOPLE'S REPUBLIC
IN THE 1980s

The microcomputing revolution is only starting.
We are at the beginning of a long road, not knowing where it will lead us,
and what will be the pace at which we will move along.[1]

Roland Wacławek

Introduction

The microcomputing revolution that Roland Wacławek[2] – an important figure for the popularisation of personal computing in Poland – refers to in the above quote is none other than the domestication of the computer. In Western countries, the computerisation of households became possible in the 1970s due to the development of microprocessor technology and the commercialisation of the idea of a personal computer. A decrease in the production cost of crucial components[3] allowed for mass production of devices that cost roughly the same as other household appliances, such as television sets and washing machines. In 1977, the opening market opportunity was quickly seized by three U.S.-made products: the Apple II, Commodore PET and TRS-80. Soon after, companies such as Atari (400/800, 1979) and Sinclair (ZX80, 1980), as well as numerous copycats around the world, joined the race. Just four decades after the construction of the first modern computer,[4] hobbyists and other curious customers could enjoy this once-futuristic technology in their own homes.

In Poland, the computerisation process began with the first locally constructed electronic computing devices, or as they were then called, "electronic brains" ("mózgi elektronowe"). The first digital machine was built between 1956–58 under the leadership of Leon Łukaszewicz at the Department of Mathematical Apparatus at the Polish Academy of Sciences, based on the existing American

---

[1] Roland Wacławek, *Z mikrokomputerem na co dzień* (Warszawa: Nasza Księgarnia, 1987), 134 (quotation translated by Maria B. Garda).

[2] Roland Wacławek, born in 1956, is an electrical engineer, press journalist and author of many popular books on personal computing and DIY electronics.

[3] For example, the affordable 8-bit MOS Technology 6502 microprocessor, introduced in 1975, played a key role in this process.

[4] For more on early computer history, see Paul E. Ceruzzi, *Computing: A Concise History* (Cambridge, MA: MIT Press, 2012).

(IBM 701) and Soviet (BESM) solutions.[5] This computer was called XYZ, and as Adam B. Empacher – a mathematician and champion of science – has pointed out, it shared some similarities with the first model of a Polish family car (so called Syrenka), also developed in the 1950s. Both prototypes had a DIY aesthetic but, as Empacher writes, even though "[XYZ] sometimes breaks, you can still conduct calculations on it."[6] However, the new computer was not only employed for its official assignment to conduct calculations for use "in science and industry,"[7] but also to play tic-tac-toe and conduct graphic experiments using an oscilloscope.[8] Although we can assume these applications were primarily experimental or demonstrative in purpose, they also expressed a very typical desire that often accompanies the introduction of a new technology: to explore its possibilities.

The emergence of a new media resulted in the rhetoric of "saddle period" ("Sattelzeit"), a concept originally developed by Reinhardt Koselleck in reference to a transformative period in German history (1750–1850), and now sometimes used to describe the general sense of breakthrough we associate with the times we live in.[9] Not dissimilar to the moment of crossing a mountain pass, when we cannot yet see what is on the other side of the ridge, so too in Wacławek's statement, we do not know what awaits us at the end of the "path to computerisation." Nevertheless, in both cases, our horizon of expectations predicts a significant qualitative change, which in this case refers to a human relationship with technology. As Wacławek recalls, during this period he had an idea about how computing technology would evolve, but he was not certain how the societal use of computers would develop.[10] In other words, although Moore's law predicted a systematic increase in computing power,[11] it was impossible to predict for what purposes this technology would be used by the users themselves. The generation that witnessed the domestication of micros also became a participant in the "microcomputing revolution," also known as the arrival of the "silicon

---

[5] Bartłomiej Kluska, *Automaty liczą. Komputery PRL* (Gdynia: Novaeres, 2013), 17–20.

[6] Adam B. Empacher, *Maszyny liczą same?* (Warszawa: Wiedza Powszechna, 1960), 118.

[7] Ibid., 115.

[8] Bartłomiej Kluska, Mariusz Rozwadowski, *Bajty polskie*, 2nd ed., revised and updated (self-pub., 2014), 7–8.

[9] Reinhardt Koselleck, *Semantyka historyczna*, trans. Wojciech Kunicki (Poznań: Wydawnictwo Poznańskie, 2001).

[10] Roland Wacławek, interview, December 4, 2016.

[11] Empirical law that assumes that "[t]he complexity for minimum component costs has increased at a rate of roughly a factor of two per year." See Gordon E. Moore, "Cramming More Components onto Integrated Circuits," *Electronics* 38, no. 8, April 19 (1965): 2.

wave" ("krzemowa fala"), a term used in the Polish press (in reference to Alvin Toffler's book *The Third Wave*[12]). It was only with time that users were able to determine what this device would be used for, as at this point the micros were still artifacts under construction.[13]

Methodology and review of the current state of research

This chapter was inspired by the works of Mikael Hård and Andrew Jamison, dealing with cultural appropriation of technology. Before we get any further, it is necessary to make a clear distinction between two notions of cultural appropriation. The first sense of cultural appropriation, that recently seems to be more popular in cultural studies, refers to the manner in which ethnic cultures are represented in media. In particular, it refers to all kinds of appropriations that involve the use of the cultural capital of one group by another, usually the more privileged one.[14] A good example of such a phenomenon is the appropriation of ethnic costumes, such as Hindu or African, by Western pop culture (e.g. in music videos). The second sense of cultural appropriation, and the one Hård and Jamison are interested in, is the process "by which novelty is brought under human control," and "new things and new ideas are made to fit into established ways of life."[15]

Hård and Jamison see "the development of technology and science as multifaceted processes of cultural appropriation."[16] In a more recent take on the subject, François Bar and his co-authors define appropriation as: "the process through which technology users go beyond mere adoption to make technology their own and to embed it within their social, economic, and political practices."[17] They also observe how Hård and Jamison's "pro-social approach focuses on three settings

---

[12] Alvin Toffler, *Trzecia fala*, trans. Ewa Woydyłło (Warszawa: Państwowy Instytut Wydawniczy, 1986). Originally published in English in 1980.

[13] See Trevor J. Pinch and Wiebe E. Bijker, "The Social Construction of Facts and Artifacts: Or How the Sociology of Science and the Sociology of Technology Might Benefit Each Other," in *The Social Construction of Technological Systems*, eds. Wiebe E. Bijker, Thomas P. Hughes and Trevor Pinch (Cambridge, MA: The MIT Press, 2012),

[14] See Beretta E. Smith-Shomade, "Appropriation," in *Keywords for Media Studies*, eds. Laurie Ouellette, Jonathan Grey (New York: New York University Press, 2017), 29–31.

[15] Mikael Hård, Andrew Jamison, *Hubris and Hybrids: A Cultural History of Technology and Science* (New York: Routledge, 2005), 4.

[16] Ibid., 13.

[17] François Bar, Matthew S. Weber, Francis Pisani, "Mobile Technology Appropriation in a Distant Mirror: Baroquization, Creolization, and Cannibalism," *New Media & Society* 18, no. 4 (2016): 1.

– the production of structures, systems, and artifacts – at three levels – discursive, institutional, and practical."[18] Such a theoretical framework allows for us to recognise a diverse range of phenomena that are important for the appropriation of microcomputers in the local and historical context.

This chapter is based on a wide variety of historical and academic sources, including oral history interviews (conducted with over 30 respondents), archival documets, an unsystematic review of computer magazines and other topical press of the era,[19] as well as a review of the growing literature on the subject. The history of computer use in the Polish People's Republic (PPR) has been studied by a growing group of researchers from various disciplines, as well as by a large group of technology enthusiasts, hobbyists associated with the retrogaming movement, and former employees of the electronic industry.

Early academic research available in English was conducted by Graeme Kirkpatrick[20] and Patryk Wasiak.[21] However, most of the existing studies were published only in Polish, including: works by Bartłomiej Kluska, who is best known for his many popular history resources on the subject[22] (some prepared in collaboration with the late Mariusz Rozwadowski[23]), and the recent anthology on electronics, computers and operational systems in the PPR, edited by Mirosław Sikora in collaboration with Piotr Fuglewicz.[24] Furthermore, there is a great number of online resources that are being collected and archived by Polish retrogaming communities and technology enthusiasts.[25]

---

[18] Ibid., 4.

[19] My intention behind the periodical review was to explore titles that were not extensively researched in previous works on the subject (e.g. by Bartłomiej Kluska and Patryk Wasiak).

[20] Graeme Kirkpatrick, "Meritums, Spectrums and Narrative Memories of 'Pre-Virtual' Computing in Cold War Europe," *The Sociological Review* 55, no. 2 (2007): 235.

[21] Patryk Wasiak, "Playing and copying: Social practices of home computer users in Poland during the 1980s," *Hacking Europe. From Computer Cultures to Demoscene*, eds. Gerard Alberts, Ruth Oldenziel (London: Springer, 2014).

[22] The prolific work of Bartłomiej Kluska, born 1980, who started publishing on the Polish video game and media history as a journalist in 2004, deserves a separate bibliographical overview and, although dispersed, remains the most wide-ranging take on the subject.

[23] Bartłomiej Kluska and Mariusz Rozwadowski. *Bajty polskie* (Łódź: Samizdat Orka, 2011).

[24] Mirosław Sikora and Piotr Fuglewicz, eds., *High-tech za żelazną kurtyną. Elektronika, komputery i systemy sterowania w PRL* (Katowice: Instytut Pamięci Narodowej, 2017).

[25] For the sake of future research, a complete meta-catalogue of these on-line materials is much needed.

Thus, case studies and materials worth using are very diverse when it comes to their form, scientific rigour, and selection of topics. Furthermore, each of these sources requires an appropriate critical approach. Let us start with publications and documents from the communist period, when Poland was an authoritarian regime and both censorship and the presence of the Security Service (Służba Bezpieczeństwa, SB[26]) were affecting everyday lives. As Wasiak notes, "a large part of press materials in Polish computer magazines, as well as articles on computers in daily newspapers and weeklies were purely aspirational."[27] Sometimes research of these materials tells us more about the ideological tensions that accompanied computerisation than about the situation itself. However, the dominance of ideology as a point of focus varies depending on the type of publication. Aspirational discourse described by Wasiak was present in popular newspapers, which were often a propaganda channel for the state, but not so much in dozens of computer manuals written by experts and hobbyists. A good example of that is the first, and allegedly most impactful computer-oriented, youth magazine of the era: *Bajtek*.[28] It was originally published as a monthly supplement to a daily newspaper – *Sztandar Młodych*, an official media outlet of Polska Zjednoczona Partia Robotnicza (PZPR, Polish United Workers' Party), and was created by its youth branch – Związek Socjalistycznej Młodzieży Polskiej (ZSMP, Polish Socialist Youth Union). As Dominika Staszenko-Chojnacka reports:

In the first issue of *Bajtek* the deputy editor-in-chief of *Odrodzenie*, Zbigniew Siedlecki, and the deputy editor-in-chief of *Sztandar Młodych*, Waldemar Siwiński, signed their names under an assurance that "the ambition of the editorial team is – very broadly speaking – fighting the microcomputer illiteracy in Poland." The brief editorial to this issue also contains a claim that school education is highly important since the role of microcomputer technology in the development of all societies is becoming evident, and "this is absolutely essential to the good future of our country."[29]

---

[26] Służba Bezpieczeństwa was the Polish secret police and chief security service of the Communist regime; an equivalent to the East German Stasi.

[27] Patryk Wasiak, "'Grali i kopiowali' – Gry komputerowe w PRL jako problem badawczy," in *Kultura popularna w Polsce w latach 1944–1989: problemy i perspektywy badawcze*, ed. Katarzyna Stańczak-Wiślicz (Warszawa: Instytut Badań Literackich PAN, 2012), 206.

[28] "Bajtek" is a humorous neologism referring both to English "byte" and to the Polish word "kajtek" ("little boy").

[29] Dominika Staszenko-Chojnacka, "Narodziny medium. Gry wideo w polskiej prasie hobbystycznej końca XX wieku" (PhD diss., Uniwersytet Łódzki, 2020), 44. The inner citation comes from Zbigniew Siedlecki, Waldemar Siwiński, "RUN czyli zaczynamy," *Bajtek*, wrzesień, 1985, 2.

However, as Wasiak notes in reference to *Bajtek*'s editorials: "[a]ppeals voiced in this tone had virtually no impact on the domestication of home computers."[30] It was simply a rhetoric one had to get used to while living in a communist country, and a rhetoric one has to be aware of while analysing these materials.

The security files available in the archives of the Institute of National Remembrance (IPN) require special attention to assess the credibility of the information included. One must remember that during interrogations by the SB, the interrogated subjects pursued their own goals (e.g. avoiding trouble), while the officers writing down the testimonies had their own agenda (e.g. proving themselves to their superiors). However, as Roman Graczyk argues, "the hypothesis of systematic falsification of records is hard to sustain… it is based on a general premise of the moral deterioration of communism: because communism was a moral evil, the records are false."[31] Graczyk observes that the SB implemented various procedures for verifying the collected information, ultimately forming the basis for decisions taken by the Party, and thus it was important that they were relatively credible.[32] Another important issue is the ethical choices researchers have to make while using sources from the IPN archives. Based on my own experience, the best practice to follow is to select material based on the best interest of the system's victims and the relevance of data to the research conducted. For example, for the purposes of this study, no personal files were used. That means I was not investigating whether a given individual was a secret service asset and collaborated with the regime or not, as it is not relevant to my study.

An important aspect of researching a period that is still present in living memory is the possibility of using oral history methods of inquiry. As Jaroslav Švelch emphasizes, one should maintain the same critical attitude towards the sources of oral history as towards any other historical sources. Nevertheless, stories relating to the use of computers in the 1980s seem to pose their own specific challenges that were aptly pointed out in the quoted presentation from Švelch, including anecdotes of unknown origin or borrowed memories.[33] I cannot stress enough how important it is to intensify this type of documentation process in the coming years. We have to remember that conducting interviews may soon become impossible due to the passing of potential respondents.

---

[30] Wasiak, "Playing and copying," 136.

[31] Roman Graczyk, *Tropem SB. Jak czytać teczki?* (Kraków: Wydawnictwo Znak, 2007), 273–274.

[32] Ibid., 273–275.

[33] Jaroslav Švelch, "Tall Tales and Murky Memories of Computer Gaming in 1980s Czechoslovakia," filmed October 18, 2019 at the 1st Collaborative Game Histories seminar in Tampere, video, 1:37:38, accessed January 30, 2020, https://youtu.be/A8OcX-wpMQcY?t=5858

The selection of interviewees was based on a review of the period literature (e.g. computer magazines), which allowed us to identify the key figures of the computerisation movement in Poland (e.g. Roland Wacławek), and was dependent on establishing cooperation with retrogaming movements (e.g. Fundacja Promocji Retroinformatyki "Dawne Komputery i Gry," Museum of Computers and Information Technology in Katowice, and Retrogralnia in Wrocław). As Wasiak notes, interviews available on retrogaming sites, which have been carried out by amateur historians over the last ten years or so, tend to "only include people who actively participated in the computer scene, creating their own programmes, graphics and music."[34] Therefore, they are a very valuable historical source, but they do not reflect the experiences of ordinary users, who were at the centre of our research interest for this book. During the research period, it also turned out that some potential respondents are already dead or are unable to be interviewed due to ill health, which only proves how important research using oral history methods that covers the 1980s, is right now, and how urgent it is that it be conducted sooner rather than later.

At this point it is worth raising the issue of the gender characteristics found within the phenomenon in question. The history of microcomputer use in the 1980s is primarily the history of young men. This does not mean that there were no women in the microcomputer scene, but it is undeniable that there were not many of them. In *Przegląd Techniczny* magazine, Adam Stawowy joked that almost everyone is interested in computers now, so soon you can expect something implicitly unthinkable – like a programming course published even in *Przyjaciółka* (a magazine for women).[35] This perception of women's relation to computer science was not only a Polish phenomenon, but unfortunately a clear and visible international trend.

Švelch considers three reasons for this situation in the context of former Czechoslovakia, which can help us to better understand similar gender tensions in the PPR. Firstly, both in the West and in the countries of the Eastern Bloc, "the computer was [...] presented and perceived as a boy's toy."[36] This was due to stereotypes that were dominant in Western culture at the time, which ascribed an interest in technology to boys rather than girls.[37] Interestingly, that did not translate into the percentage of women studying computer science, of whom there

---

[34] Wasiak, "'Grali i kopiowali," 212.

[35] Adam Stawowy, "Komputery w instytucjach," *Przegląd Techniczny*, no. 1 (1987): 13.

[36] Jaroslav Švelch, *Gaming the Iron Curtain: How Teenagers and Amateurs in Communist Czechoslovakia Claimed the Medium of Computer Games* (Cambridge, MA: MIT Press, 2018), 78.

[37] For more on gender and computer technologies, see Sherry Turkle, *The Second Self: Computers and the Human Spirit* (Cambridge, MA: MIT Press, 2005); Jane

were many in the 1970s. In addition, women were an important workforce when it came to operating and producing mainframe computers (e.g. at the Elwro plant in Wrocław).[38] Furthermore, the hobby movements that later grew into micro-computer clubs, such as radio amateurs, were traditionally masculinised. For example, in the Czechoslovak Svazarm (a paramilitary organisation), on average nine out of ten members were male. Men also simply had more time to engage in entertainment, including various hobbies, because of various social structures in place. As Švelch notes, "[d]espite several progressive gender policies, Czechoslo-vak society was a patriarchal one, and a relative gender equality in labour did not bring forth gender equality in leisure."[39]

The situation was similar in Poland. During the Stalinist period (1948–56), one of the main Party slogans was "women's empowerment," and in fact a signif-icant number of women took up jobs in professions previously reserved only for men (another famous slogan from that era was "Women, to tractors!"). However, after the fall of the Stalinist doctrine in the second half of the 1950s, criticism lev-eraged at the renewed Party indicated that women's empowerment was not feasi-ble by means of social change, and Stalinist experiments proved to be detrimen-tal to the functioning of the family. Following this, the Party was only interested in the role of women as mothers, while trying, at least at the level of propaganda, to help them combine parental and professional responsibilities.[40]

Based on these sources, an attempt can be made to recreate not only the process of domestication of microcomputers in Poland in the 1980s, but also the accompany-ing emotions and cultural practices. Although the respondents usually talk about their specific individual experiences, scenarios and anecdotes that reoccur and resemble each other allow us to guess which practices associated with the adaptation of this technology were typical for the experience of using microcomputers in the late PPR. That does not mean, however, that this study claims to be exhaustive. Furthermore, there are still many contexts that require separate discussion, from the aforemen-tioned history of women in the Polish computer science of the 1980s,[41] to unusual or marginal phenomena; some of which I discuss in the conclusion.

Margolis, Allan Fisher, *Unlocking the Clubhouse: Women in Computing* (Cambridge, MA: MIT Press, 2002).

[38] It can be observed in the newsreels from the Elwro manufacturing plant from that time.

[39] Švelch, *Gaming the Iron Curtain*, 79.

[40] See Małgorzata Fidelis, *Women, Communism, and Industrialization in Postwar Poland* (Cambridge: Cambridge University Press, 2010), 202.

[41] For the most recent take on the topic (in Polish) that unfortunately came out too late to be included in this book, see Karolina Wasilewska, *Cyfrodziewczyny. Pionierki polskiej informatyki* (Warszawa: Wydawnictwo Krytyki Politycznej, 2020).

On two sides of the Iron Curtain

In the 1980s, a Polish television science programme called *Sonda* (*Probe*) had an unprecedented role in national science education. Watching the programme hosted by two charismatic science educators (as we would call them today), Zdzisław Kamiński and Andrzej Kurek, was a generational experience, and not only for the youngest. Marcin Borkowski[42] recalls how he was always in a hurry after work so that he did not miss *Sonda* at 6 pm.[43] Due to the general poor quality of content presented on the only two available national television channels, *Sonda* was a phenomenon, one followed and commented on by almost everyone with an interest in technology.

Czynnik Si (*The Si Factor*) was a limited series broadcast in the summer of 1984, as a *Sonda* special. Like the entire *Sonda* series, it had a very large viewership. The first episode discussed the history of microchip technology, telling the story of the pioneers and presenting the American Silicon Valley as a kind of forge of the future. Kamiński and Kurek tried to answer questions, such as what the purpose of this new technology was, and whether microchips were a toy, a tool or a threat. The presenters also explored the many possibilities of human life improvement, but the audience was also warned about the threats of automation. The narrator reported that "the silicon deluge will result in uninhabited islands of technology,"[44] referring to, for instance, automated manufacturing plants. In a country ruled by the Polish United Workers' Party, such visions had to cause significant concerns.

It should be noted that *Sonda* was often assembled from materials that the programme editors received thanks to private international connections. The first episode of *Czynnik Si* was based on television materials obtained from abroad, namely the documentary *Now the Chips Are Down* (BBC, 1978). According to British researchers, this programme played an important role in popularising computers in the United Kingdom.[45] It is interesting to see how the same visual material was conceptualised between the original and the Polish version, for which a new voiceover was recorded. I would like to draw attention to a brief, almost half-minute segment dealing with arcade games as a metaphor for the Cold War (23.30-24 min.). The segment starts with footage of young people playing video games such as *Guided*

---

[42] Marcin "Borek" Borkowski, born in 1962, is a computer programmer, press journalist dealing with computer games, and creator of one of the first Polish text adventure games *Puszka Pandory* [*Pandora's Box*] (ZX Spectrum, 1986).

[43] Marcin Borkowski, interview, December 4, 2016.

[44] *Sonda, Czynnik Si special*, part 3, min. 8:05.

[45] See Alison Gazzard, *Now the Chips Are Down: The BBC Micro* (Cambridge, MA: MIT Press, 2016).

Missile (*Midway Games*, 1977) at an arcade.[46] This image is juxtaposed with a salvo of US ground-to-air rockets, followed by a close-up of a Soviet bomber dropping lethal cargo. The British commentator reports that although video games may seem trivial, it is a growing business that already brings a quarter of a million pounds of income annually. Furthermore, the technology itself was developed due to demand from the US military, which uses it in advanced defense systems. The Polish commentator mentions that microprocessors are used by hostile armies, but his voiceover creates a completely different narrative, one in which arcade games are seemingly "frivolous toys" for the Western consumer, while actually serving as an instrument of capitalist indoctrination. The free market aspect of video games' financial success was omitted, and the image of the Soviet bomber was edited out, delivering a very different message.

Graeme Kirkpatrick points out that on both sides of the Iron Curtain, despite major system differences, computers were embroiled in contradictory ideological messages.[47] In the same piece, Kirkpatrick notes that in the Eastern Bloc countries during the 1960s and 70s, computers were a very important tool of propaganda, and their role in the pursuit of the ultimate communist society was often emphasised. Yet access to them, as well as to other technologies related to information processing, was strictly regulated.[48] I would say that they often fulfilled this role, at least in theory, yet the premise of this chapter is to show that users were able to find alternative ways to interact with the technology. Kirkpatrick further elaborates that, in capitalist countries, computers were presented alternatively as an announcement of a future free from labour, and as a real threat to civil liberties. To further problematise the inconsistency of this message, Kirkpatrick notes that some Soviet propagators of cybernetics perceived themselves as dissidents, while early American hackers saw the computer as a tool of libertarian socialism.[49] I agree with Kirkpatrick that discourses around new technologies rarely, if ever, offer a coherent message. As it turns out, even in the context of the Eastern Bloc countries, despite many systemic similarities, one can still speak of "national styles and appropriation trajectories."[50]

---

[46] On the Military-Entertainment Complex, see Tim Lenoir, Henry Lowood, "Theaters of War: The Military-Entertainment Complex," in *Collection – Laboratory – Theater: Scenes of Knowledge in the 17th Century*, eds. Helmar Schramm, Ludger Schwarte, and Jan Lazardzig (Berlin: Walter de Gruyter, 2005).

[47] Kirkpatrick, "Meritums, Spectrums," 235.

[48] Ibid.

[49] Ibid. See also Richard Barbrook, Andy Cameron, "The Californian Ideology," *Mute* 1, no. 3 (1995), accessed May 21, 2020, http://www.metamute.org/editorial/articles/californian-ideology

[50] Hård, Jamison, *Hubris and Hybrids*, 164.

Top-down and bottom-up computerisation

Top-down computerisation and the technological gap

In the 1980s, Polish public discourse on computers often debated the technological gap[51] that the country was struggling with: in terms of technical development, Poland was significantly lagging behind the West. The key factor at work here was the Coordinating Committee for Multilateral Export Controls (CoCom), i.e. an embargo on new technologies that covered all countries of the Eastern Bloc.[52] It was introduced by the United States and several other Western countries in the late 1940s, at the very dawn of the Cold War. This embargo was meant to prevent the transfer of technologies that could have been used in a military context. In 1949, an equivalent international organisation – the Council for Mutual Economic Assistance (Comecon) – was established to coordinate the economic cooperation of the Eastern Bloc countries.

Due to CoCom, the Polish computer industry had to rely on domestic suppliers, or components coming from other Comecon countries. Interestingly, the Eastern Bloc's technological lag in the area of computing did not occur linearly. Initial outdatedness, resulting from the fact that modern computers were shaped by technology developed in Western countries, began to decrease in the 1960s. As Kulisiewicz estimates, the lag behind Western releases had decreased from seven years in the early 1950s to just four in the mid-1960s.[53] Nevertheless, as early as the 1970s, alongside the commercialisation of microprocessor technology in the West, the Eastern Bloc began to regress again. As a result, in the early 1980s, Poland was dependent on the West when it came to importing components used in modern computers.[54] In 1985, the editor of *Przegląd Techniczny*

---

[51] Technological gap theory is a model developed by economists in the early 1960s that deals with the issue of dynamic distance between developed and developing countries. See Michael V. Posner, "International Trade and Technical Change," *Oxford Economic Papers* 13, no. 3 (1961).

[52] See Michael Mastanduno, *Economic Containment: CoCom and the Politics of East-West Trade* (Cornell University Press, 1992).

[53] Tomasz Kulisiewicz, "Polskie komputery 1948–1989. Produkcja i zastosowania na tle geopolitycznym i gospodarczym," in *High-tech za żelazną kurtyną. Elektronika, komputery i systemy sterowania w PRL*, ed. Mirosław Sikora with Piotr Fuglewicz (Katowice: Instytut Pamięci Narodowej, 2017), 45.

[54] See Kulisiewicz, "Polskie komputery," 57. See also Władysław Majewski, "Z czego lepić komputery? Rozmowa z Henrykiem Piłko," *Przegląd Techniczny*, no. 10 (1985): 10.

commented harshly that Poland was "a country which in the field of modern integrated circuits production limits itself to giving away state awards..."[55] To understand the sarcastic undertone of this comment we have to acknowledge the dual realities of everyday life in a communist state, where the Party's official narrative (represented above by the state recognition of the industry's alleged achievements) could have been in direct contradiction with the facts (the actual lack of components). In short, what would be considered an objective failure could have always been recognised as a success, if the political *status quo* and the Party's best interests demanded it.

Nonetheless, for the communist authorities, this situation was quite problematic. Within the Party's doctrine, so-called scientific and technical progress was said to constitute an important means for the ultimate liberation of the proletariat. According to Švelch, the technical staff experienced a rather limited repression during the purges that followed the Prague Spring of 1968, proving that the regime did not perceive technology as a tool of ideological struggle.[56] It seems that across the entire Eastern bloc, attitudes towards technology were definitely favourable. For practical and ideological reasons, all modern technology was good and progressive. It seems that in other European countries, even semi-peripheral nations such as Finland, as well as in the US, it was the case that "scientists and other actors perceived the computer as part of modern progression, which had to inevitably be adapted sooner or later, one way or another."[57] It is, however, worth recognising the sheer scope and idealistic nature of communist initiatives, at least at the intentional level, and how this decisively distinguished them from Western approaches. The proposed projects were truly massive in scale and were supposed to ensure access to new technologies as another public service, just like sewage or public transportation systems.

However, it should not be forgotten that the authoritarian communist states, focused as they were on central planning, tended to control all manifestations of social life, treating them as a means to achieve assumed goals and stages of development. As Vítězslav Sommer writes about these relations in former Czechoslovakia, "[s]cience was not an active actor of development but a policy instrument constructed by experts for the purposes of centralised governance by the party elites."[58] In the 1960s, a utopian vision

---

[55] W.M., *Przegląd Techniczny*, no. 42 (1985): 21.

[56] Švelch, *Gaming the Iron Curtain*, 5.

[57] Jaakko Suominen, Jussi Parikka, "Sublimated Attractions: The Introduction of Early Computers in Finland in the Late 1950s as a Mediated Experience," *Media History* 16, no. 3 (2010): 323.

[58] Vítězslav Sommer, "Scientists of the World, Unite! Radovan Richta's Theory of Scientific and Technological Revolution," in *Science Studies during the Cold War*

of scientific and technical revolution was developed in Czechoslovakia, but what it amounted to, was the computerisation of scientific research and the production of goods in state-owned enterprises, rather than bringing computers into homes.[59] Švelch notes that in the socialist countries of the Eastern Bloc, the official state discourse focused on communal rather than individual access to technology.[60] Obviously, this was associated with the traditional line of communist party policy: the rejection of all kinds of elitism. It seems that in comparison to Czechoslovakia, the official line of the Polish communist party placed computing technologies much higher on the list of priorities, something that Kirkpatrick has already speculated about,[61] and which Kluska's research has confirmed.[62]

The authorities of the PPR, during both the administrations of Władysław Gomułka (1956–70) and Edward Gierek (1970–80), were vitally interested in the possibilities of computer technology. Various types of government positions and committees were established to facilitate not only the computerisation of the country, but also the computerisation of society.[63] Ambitious goals were set, such as the establishment of KSI (National Information System), which was responsible for "covering all aspects of central state control with the network of electronic computing technology."[64] It is not difficult to imagine that for all communist governments, not only those in the PPR, computers brought hope for the large-scale implementation of the command-and-control economy model.[65] Let us not forget that already in the 1950s, the Main Computer Centre of the State Planning Committee of the USSR (so called Gosplan) was established, "for the purpose of introducing electronic computers into the practice of planning and economic calculations"[66] on a national level. But regardless of the legitimacy of such macroeconomic concepts, their feasibility

---

and Beyond, eds. Elena Aronova, Simone Turchetti (Basingstoke: Palgrave Macmillan, 2016), 193–194.

[59] See Švelch, Gaming the Iron Curtain, 10–12 and 28.

[60] See, ibid., 34.

[61] Kirkpatrick, "Meritums, Spectrums," 232.

[62] Kluska, Automaty liczą, 125.

[63] Among others, the Office of the Government Plenipotentiary for Electronic Computing Technology (1964–1972) and the National Bureau of Information Technology (1971–1975).

[64] Kluska, Automaty liczą, 111. See also Kulisiewicz, "Polskie komputery," 50–51.

[65] See Kulisiewicz, "Polskie komputery," 42.

[66] Vladimir Kitov, Nikolay Krotov, "The Main Computer Center of the USSR State Planning Committee (MCC of Gosplan)," Selected Papers: 2017 Fourth International Conference on Computer Technology in Russia and in the Former Soviet Union (SORUCOM), Zelenograd, 2017, 228.

was simply impossible under the conditions of the PPR. As Kulisiewicz observes, the functioning of the nationalised, centralised economy was based on a complex system of relationships and tensions between individual production facilities and power centres at various levels across respective industry sectors. The so-called "bargaining with the centre," meaning multilevel negotiations on "indicators and plans, and the allocation of resources,"[67] often led to falsification of production results. Thus, even if a computer model of the economic sector had been developed, the data used by this projection would not have been reliable, undermining the entire undertaking.[68] The only major project that was successfully implemented, at least to some extent, during that period, was the national identification number system (so called PESEL).[69] Given the potential importance of the identification system for state surveillance, as well as the direct supervision of the PESEL project by the Ministry of the Interior, to which the SB was subordinate, one can ask a rhetorical question: what was the priority for the authorities?

Computers by the Vistula River

According to Švelch, in Czechoslovakia, "[m]icrocomputers for the home or personal use never made it into the five-year plans of the state socialist economy."[70] In this respect, the authorities of the PZPR were more determined than their southern neighbours, as such plans existed and were even partially implemented in the 1980s. However, representatives of state institutions and related organisations were often not convinced that the idea of a computer meant for personal use was a priority. Established in 1986, Ogólnopolska Fundacja Edukacji Komputerowej (the Polish National Computer Education Foundation) which was an association of key organisations related to the country's informatisation and based in Wrocław, predicated that "before the computer finds its way into everyday homes, it must find itself in schools."[71] This reflects the official government stance and explains later efforts to bring computers to schools. The foundation's purpose was the long-term IT education of citizens (mainly in programming), starting with familiarising future users with computers.

---

[67] Kulisiewicz, "Polskie komputery," 51.

[68] See ibid., 50–51.

[69] See Bartłomiej Kluska, *PESEL w PRL: informacja czy inwigilacja* (Łódź: Księży Młyn Dom Wydawniczy, 2019).

[70] Švelch, *Gaming the Iron Curtain*, 2.

[71] *Motywy*, May 20, 1987: 2.

Fig. 1. Computer room equipped with Mazovia computers, from the film *Pan Kleks w kosmosie* (*Mr. Blot in Space*, Krzysztof Gradowski, 1988)

Source: National Film Archive

To achieve these goals, computers made in Poland were to be used. In this context, three devices should be mentioned: Meritum (compatible with TRS-80), Elwro Junior (compatible with ZX Spectrum) and Mazovia (a clone of IBM PC).[72] However, due to the small scale of production, none of them had a significant impact on the computerisation of the PPR. For example, Mazovia is probably most recognisable today as a prop used in a science-fiction movie for young audiences, *Pan Kleks w kosmosie* (*Mr. Blot in Space*, Krzysztof Gradowski,

---

[72] More about the history of Polish microcomputers: Kluska, *Automaty liczą*, 146–160 and 174–195.

1988). This cinematic vision of the future of Polish education served as a metaphor for the utopian concepts of computerisation that came to life in the Eastern Bloc countries. Although utopian, these initiatives can be appreciated for their planned scope and pro-social agenda to democratise access to new technologies. In the end, however, these projects turned out to be only thought experiments that came closest to realisation as part of a science-fiction movie.

Yet perhaps the most interesting story is that of the Junior, a device "designed by a team from the Poznań University of Technology headed by Wojciech Cellary."[73] The machine produced by WZE Elwro in Wrocław was well adapted to Polish realities, because, according to Kluska, it was characterised by "high tolerance… to voltage fluctuations in the network."[74] It is also worth mentioning that various pre-existing components were used in order to speed up the production process. This included a toy musical keyboard (Elwirka), used as the machine's casing, thanks to which the computer came with a stand intended for sheet music. As Sysło reports, "10,000 units were produced, which equipped about 1,000 schools in Poland."[75] Wojciech Cellary, Junior's chief designer, put forward an optimistic hypothesis that, thanks to his computer, "in the second half of the 1980s, one million Polish students… first came into contact with a computer."[76] However, one should remember that Cellary's estimations were based on an assumption of the rotational use of the machine by numerous pupils during one lesson. Such a short interaction with the computer would probably engender some initial familiarisation with the technology, but did not allow students to develop the skills necessary to operate the device at an advanced level. The proposed rotational use of computers could have been a bureaucratic strategy to overcome equipment shortages, as it also appears in other documents related to the computerisation of education from that period. For example, the governmental Programme for the Development of Computerisation of Academic Institutions assumed that in 1995, one computer would be used by 15 students in the field of humanities.[77] Cellary writes, probably in accordance with the original intent behind

---

[73] Maciej M. Sysło, "Zasługi PRL dla edukacji informatycznej," in *High-tech za żelazną kurtyną*, 353.

[74] Kluska, *Automaty liczą*, 178.

[75] Sysło, "Zasługi PRL," 353.

[76] Wojciech Cellary, Paweł Krzysztofiak, "Historia polskiego komputera edukacyjnego," December 22, 2015, accessed January 30, 2020, www.computerworld.pl/news/404071_1/Historia.polskiego.komputera.edukacyjnego.html

[77] Ministerstwo Edukacji Narodowej, *Projekt resortowego programu badawczo-rozwojowego RRI.14 "Informatyzacja procesów dydaktycznych i naukowo-badawczych w szkołach wyższych"* (Wrocław: 1998).

its creation, that the Junior computer "prepared the country for the IT revolution."[78] Yet the truth seems rather different. When this native microcomputer was introduced to schools in 1987, allowing students for the prescribed rotational use, the domestication of computers was already in full swing as a result of bottom-up hobby movements.

Microcomputers, move out!

There is no scientific consensus on the scale of personal computer diffusion in Poland. Due to the lack of statistical data from the earliest period of the process (the 1970s and early 1980s), as well as the multitude of distribution channels that operated in the shadow economies, it is not possible to obtain accurate figures. Barbara Łukasik-Makowska estimated in a 1983 issue of *Przegląd Techniczny* that there were about one thousand to three thousand microcomputers in Poland.[79] Moreover, as reported by Bartłomiej Kluska, "in a study conducted in the summer of 1988 by CBOS, already 3% of Poles boasted of having a computer at home (for comparison: 35% of those interviewed admitted to having a colour TV, 4% a VCR, 1% a compact disc player, and 0.6% a satellite TV tuner)."[80] The situation in Poland did not differ from other countries of the Socialist Bloc. In 1989 Czechoslovakia, only 1.8% of households had a computer, while in the United Kingdom it was 18%.[81] These devices came to the country mainly thanks to a grassroots initiative. Agnieszka Wróblewska estimated that in the mid-1980s, approximately 20-30,000 new home microcomputers (purchased with private funds) were appearing on the Polish market every year.[82] We can only state with certainty that there was an increase in the number of microcomputers following the mid-1980s. The results of the aforementioned CBOS survey allowed Kluska to estimate the number of microcomputers in Poland at approximately 810,000.[83] According to Sysło, the number of locally produced microcomputers was close to 20,000 units (almost 10,000 units of Meritum and a similar number of Elwro

---

[78] Cellary, Krzysztofiak, "Historia polskiego komputera edukacyjnego."

[79] Barbara Łukasik-Makowska, "Sprzężenie zwrotne: Mikrokomputery, wystąp!," *Przegląd Techniczny*, no. 20 (1985): 22.

[80] Bartłomiej Kluska, "'Komputeryzacja jakby od końca' obywateli, przedsiębiorstw i uczelni PRL-u," in *High-tech za żelazną kurtyną*, 389; the referenced survey was published in Andrzej Florczyk, "Ile komputerów jest w naszych domach?," *Komputer*, no. 1 (1989): 5.

[81] Švelch, *Gaming the Iron Curtain*, 1.

[82] Agnieszka Wróblewska, "Bilion do podziału," *Przegląd Techniczny*, no. 45 (1986): 8.

[83] Kluska, "Komputeryzacja jakby od końca," 389.

Juniors).[84] These estimations leave no place for speculation: the domestication of computers took place because of the machines imported from outside of Poland. In 1987, when Junior found its way to schools, many educational and cultural institutions were already equipped with other brands of microcomputer. Furthermore, those living in urban centres were likely to be friends with at least one early adopter. It is true that it was usually a very simple device, such as ZX Spectrum or Atari 800 XL, which served mainly as a gaming platform, and that they were often second-hand. But gameplay being the first interaction with a computer does not undermine the experience itself.

Bottom-down computerisation and the cultural gap

As early as 1985, Mirosław Bieszki and Marian Pianowski were already expressing their concern that the technological gap associated with computers could turn into a cultural gap.[85] After all, a lack of access to the latest media platforms also meant a lack of access to cultural productions published on them. It is worth remembering that computers offered contact not only with video games, but also with other cultural productions. For example, Piotr Fuglewicz imported classic Russian literature from the USSR (including some written by dissident authors) on memory disks, and then printed it in Poland.[86] Tomasz Cieślewicz recalls that he and his friend were playing demoscene productions just to hear samples from the latest Michael Jackson song.[87] During this time, cultural works recognised by the state censorship system as "safe," (which we would today call '"politically correct"), were relatively available on the market. However, there was a growing desire to consume "forbidden" Western culture: the books, movies or other works that were rejected on the grounds of censorship and which were not in official circulation. The underground publishing and distribution of cultural works was quite a common practice. Access paths to cultural works published in the West were created by grassroot movements, and ways of accessing computer technologies were developed in a similar way.

As Kluska observes, in the 1970s, an earlier image of a computer, seen as "[a] mysterious *electronic brain*, flashing with many lights, and playing chess or controlling space vehicles," was replaced in the public consciousness by "a useful tool liberating man from the ballast of routine, mechanical and tedious mental

---

[84] Sysło, "Zasługi PRL," 353.
[85] Mirosław Bieszki, Marian Pianowski, "Luka technologiczna," *Przegląd Techniczny*, no. 44 (1985): 28–29.
[86] Piotr Fuglewicz, interview, November 23, 2015.
[87] Tomasz Cieślewicz, interview, November 16, 2015.

work."[88] Unfortunately, already in the early 1980s a rather different belief began to dominate in the Polish public discourse: the belief that "computers are a distant abstraction, devoid of any impact on everyday life in the PPR."[89] To make matters worse, according to Kluska, any errors in government bureaucracy were often explained in the press as "a computer error" and as such "these machines, and their alleged mistakes, became a convenient excuse for the officials."[90] Social moods of the time are perfectly, albeit sarcastically, reflected in the following fragment of dialogue from the cult Polish comedy film *Teddy Bear* (*Miś*, 1980). The scene takes place in the Warsaw apartment of the movie's heroine (Ola), when an official (Collector) comes to collect money for an energy bill. In the dialogue, Ola greets the Collector, who then tells Ola's guest (Paluch) about his experiences with using a computer at work as they discuss measuring energy usage:

Collector: Merry Christmas.
Ola: Thank you.
Collector: Let me see what it says that you owe here.
Paluch: So, to know how much to collect and from whom, you must be pretty on top of everything, right?
Collector: It used to be like that, but now we have a computer.
Paluch: Computer?
Collector: Yes, but it will always make mistakes when adding, sir. Not a month has gone by without some error being made.
Paluch: But now you don't have to know so much about your job?
Collector: Now I don't. It's much easier now, sir.

As Ewa Mazierska writes about the director of *Miś* (*Teddy Bear*, Stanisław Bareja, 1980), "[his] principal methods were exaggeration, intensification and incongruous juxtaposition."[91] Bareja's craftsmanship and the success of his comedies lie in the ability to keep the above-mentioned exaggeration in the realm of possibility. Although the Collector's view on computers seems somewhat absurd, the scene most likely felt real to the audiences, and it is likely to have been even inspired by a real opinion, or a bit of an overheard conversation.

In the face of economic struggles and shortages during the late communist era, which manifested as a daily deficit of basic supplies and empty shelves in groceries stores, a "commonsense" belief spread among parts of Polish society, which

---

[88] Kluska, *Automaty liczą*, 125.
[89] Ibid., 141.
[90] See ibid.
[91] Ewa Mazierska, "The Politics of Space in Polish Communist Cinema," in *Via Transversa: Lost Cinema of the Former Eastern Bloc*, eds. Eva Näripea, Andreas Trossek (Tallinn: Eesti Kunstiakadeemia), 231.

is reflected in the following statement by Krystyna Zielińska, a well-known radio personality and at that time also a member of the Polish parliament: "[f]irst, let's deal with bread and washing powder, with dirty toilets and bad public transport, and then the time will come for electronics, robotisation and computerisation."[92] However, as commentators from the specialist press agreed, this kind of approach would be disadvantageous in the long term.[93] It is worth examining the history of media in other developing countries that have eventually narrowed the technology gap. In India, for example, traditional telephone lines are not being developed because the process of innovation diffusion has skipped a stage, and Indian society adopts mobile technologies much faster than the landlines still lagging in infrastructure.[94] The diffusion of innovation does not have to be a linear process, and its local manifestations do not have to duplicate the scenarios known from earlier cases, in particular those taking place in different cultural realities.

As Roman Dawidson[95] noticed concerning the breakthrough of computerisation in Poland:

> [...] step by step personal computers are leaving the warmth of family homes behind and are beginning to appear in various public institutions. Such a path of development is probably specific only to our country. Under normal circumstances this process would go the other way round. People first met with computing devices at work, in large offices, banks, etc. The computer was a known machine, though still a bit mysterious... But in Poland there is a different phenomenon. It is the computerisation of the country that happens from the bottom-up. The youngest part of our society is doing this.[96]

The bottom-up computerisation, which Dawidson draws attention to and whose main driver was, in fact, the hobbyist movement, developed in parallel with top-down computerisation, which was controlled by the state. Both processes overlapped, and in the case of microcomputers it is difficult to speak about one direction of diffusion of innovation, either top-down or bottom-up, without the other. As a result, at the end of the 1980s, there were two co-existing visions of computing culture in Poland. On the one hand, there was institutionalised

---

[92] Krystyna Zielińska as paraphrased in Jacek Świdziński, "Krzemowe wyzwanie," *Związkowiec. Tygodnik Popularny*, July 6, 1986, 1.

[93] See j.r., "Komputeryzujemy się," *Komputer*, no. 8 (1986): 4–5.

[94] See Sanjay K. Singh, "The Diffusion of Mobile Phones in India," *Telecommunications Policy* 32, no. 9–10 (2008).

[95] Roman Dawidson, who died in 2011, was the head of the Technological Progress Department in *Przegląd Techniczny*, later one of the presidents of the Polish National Foundation of Computer Education (OFEK).

[96] R.D. [Roman Dawidson], *Przegląd Techniczny*, no. 40 (1985): 21.

computerisation, focused on state institutions (including school) and enterprises of various sectors. On the other hand, there was spontaneous computerisation, gathered around computer fairs, to be addressed in the final sections of this chapter. Interestingly, both computerisation cultures often functioned in the same space. From Monday to Friday, the school was a place of IT education, while on the weekends a computer fair was organised in the same building.

Breakthrough of domestication

The breakthrough in the domestication of computers in Poland took place in the mid-1980s, most likely between 1984 and 1986. In the global context, this might have been relatively late, but in the context of the Eastern bloc it seems that Poland was within the norm. There are two main reasons behind this chronology: one international, one local. Firstly, on an international level, the embargo on 8-bit technology was relaxed in 1984. Computers had been at the heart of the CoCom debate since the mid-1970s, but – as Mastanduno reports – it was not until July 1984 that the embargo on the most popular 8-bit microcomputers was removed, even though at the same time new restrictions were introduced regarding various telecommunications software and solutions.[97] Secondly, on a local level, as Kluska reports, in the autumn of 1984, the "[Polish] customs office ceased to make it difficult for citizens to import microcomputer equipment."[98] That last reason, as Kluska notes, was of more importance for the average commercial tourist than the less known and constantly evolving regulations related to CoCom, the interpretation of which depended largely on the specific situation. For example, Kluska brings to light the story of Mr. Przemysław, who is described in an issue of *Informatyka* magazine.[99] In April 1984, Mr. Przemysław received a package from his brother in Toronto containing a VIC-20 computer, peripheral devices and a set of cassettes with software (such as English learning programmes and video games). However, the customs office in the Polish seaport of Gdynia seized the package, claiming that the computer was not essential to the professional and academic work undertaken by Mr. Przemysław. According to Kluska, the story had a happy ending and after four months (August 1984) the decision was overturned and the computer was released.[100] The timeline of the events seems to suggest a possible correlation with the CoCom revision (July 1984),

---

[97] Mastanduno, *Economic Containment*, 269.
[98] Kluska, "Komputeryzacja jakby od końca," 381.
[99] Ibid., 386.
[100] Ibid.

but a more detailed investigation and solid archival proof would be required to validate this connection.

Another observation I have made corroborates this chronological hypothesis. From 1985 onwards, there was a significant increase in computer-related publications. Importantly, this new wave of publications approached microcomputing not as part of a futuristic narrative, describing technology that may one day be available, but rather in the form of practical tips for users who owned computers at home. As Kluska reports, numerous periodicals on this topic were created during this time, and it seems that 1986 was a year especially abundant in new titles:

> Newspaper stands of the time carried the following: a supplement to the *Żołnierz Wolności* titled *IKS* (*Information Technology – Computers-Systems*); *Mikroklan* which became a standalone title at last (previously it was a supplement to *Information Technology*) and was printed in colour in Vienna on good quality paper; *Informik* which evolved from a column in *Młody Technik*; and most importantly *Komputer*, the information technology culture magazine: "for mature readers."[101]

Based on Dominika Staszenko-Chojnacka's research, we can say that computer magazines of the late 1980s were focusing on educational aspects of computing and information technology, as the editorial teams were trying to align with the state narrative on the topic, e.g. video games were marginalised, even though the editors were aware that this is what really attracts kids to the microcomputers.[102]

Not only standalone titles should be taken into account here, but also numerous additions and regular columns in many magazines of various profiles. Examples include "ABC komputerów" in *Świat Młodych*, a magazine related to the Polish scouting movement and known for its Polish comic strips section, and *TOP*, a magazine interested in the growing consumer market, and a foretoken of the upcoming changes in the Polish economy.

Finally, dozens of popular science books and user guides were published, discussing the use of a computer in a variety of purposes and contexts, from professional to recreational. Some of these were simply manuals of various microcomputers translated into Polish, while others were more ambitious, like Roland Wacławek's *Mikrokomputer na co dzień* (*Everyday Living with a Microcomputer*) which will be discussed further on. Before we get to the actual uses of these machines, I would like to analyse in more detail the process of diffusion itself.

---

[101]   Kluska, *Automaty liczą*, 163; the internal quote comes from *Komputer*, no. 1 (1986): 1.

[102]   Staszenko-Chojnacka, *Narodziny medium*, 46.

First contact

Until the 1980s, few people in Poland had direct contact with modern computers. As I have previously mentioned, the history of Polish computers dates back to the 1950s, and the mass production of such devices only fully developed in what is known as the Gierek decade (1970–1980). Moreover, large calculating machines – devices such as the Odra series or RIAD computer systems, were used primarily in research institutions and for industrial purposes. There were, however, people who had contact with a computer in other contexts: for example, at their workplace or through family members. One such interesting case is the story of Marcin Borkowski, who, due to his mother's profession, was able to interact with the latest digital technologies as a young boy. His mother, EngD Barbara Borkowska, was an energy engineer involved in forecasting loads in the electric networks, and she worked at the Centre for Computerisation of Energy and Atomic Energy in Warsaw. Borkowski mentioned that in his home, wasted perforated cards used to programme mainframe computers were reused for taking notes or making shopping lists. Before he even learned the secrets of programming and could grasp the essence of the tape punching system, young Marcin assembled various words and slogans from these patterns.

As Marcin recalls, one day Barbara took him to the headquarters of the National Power Dispatch Centre where Control Data computers from the CDC 300081 series were used.[103] The powerful device occupied the entire room, and was isolated from the rest of the building by double glass doors, only for computer operators to pass through. It is worth remembering that people ordering calculations using large computing machines were often not even in their vicinity. As Borkowski himself emphasises, "such people as me, who saw this computer through this glass door, were a fraction of society."[104] So where did the awareness of the functions and importance of computers come from? After all, many of our respondents, even the youngest ones, remember their first contact with a computer as a great experience, eliciting strong emotions that have not faded with time. This means that even young children in the PPR had some concept of what a computer was, even if they did not have firsthand experience. Interestingly, only a few interviewees recalled reading a book or watching a film as their first memory related to a computer device. For example, Tomasz Cieślewicz remembers

---

[103] See Andrzej Kłos, "Rys historyczny rozwoju informatyki w polskiej elektroenergetyce," paper presented at the conference "50 lat zastosowań informatyki w polskiej energetyce," Warsaw, April 21, 2009, accessed January 30, 2020, http://apw.ee.pw.edu.pl/sep-ow/PLI/konf/zipe'09/klos/RysHist-InfwEE-AK.htm

[104] Marcin Borkowski, interview.

that the film *Tron* (Steven Lisberger, 1982) which he had seen on VHS, made a big impression on him, but it is difficult to say if he had seen it before he saw a computer in real life.[105]

The notion of the sublime is the second most important aesthetic category next to beauty, yet due to the lack of a colloquial equivalent, it is not used so often. The sublime is often connected to the experience of grandeur, as well as the splendour of nature, as for example in descriptions of natural wonders known from romantic poetry, such as high waterfalls or majestic mountain ranges. However, in media research, beginning with Leo Marx[106] but predominantly in the works of David Nye,[107] it also refers to technology. In this approach, sublimity "simply means the joy related to the triumph of reason over its surroundings, over the environment in which it operates, regardless of whether said environment is natural or not."[108] In the nineteenth century, such a sensation was often experienced while travelling by railway or encountering large scale electric illuminations. In the twentieth century, the technological sublime became associated with skyscrapers, and later with space rocket launches, among other examples. Today, similar impressions are evoked by reports of spacecraft landings on Mars or, more recently, New Year's Eve drone shows, when swarms of flying machines glide through the night sky in compact formations. Public showings of new technologies are often a source of technological sublime, and reports from these events reach a wide audience, awakening an idea of the next big thing and the desire to experience the *zeitgeist*.

Marcin Borkowski recalls how, at an exhibition organised as part of the Children's Day celebration in 1975 at the Palace of Culture and Science in Warsaw, he had the opportunity to directly interact with a domestically produced computer for the first time in his life (unfortunately we were unable to determine the model). As in the history of one of the first computer games, *Tennis for Two* (Higginbotham, 1958), here also the operators of the device decided to do something special for the young attendees. Visitors could play a geographical quiz on the machine: the programme guessed what country the player was thinking about, asking questions that simultaneously checked their general knowledge.

---

[105] Tomasz Cieślewicz, interview.

[106] Leo Marx, *The Machine in the Garden: Technology and the Pastoral Ideal in America* (New York: Oxford University Press, 1964), 198.

[107] David E. Nye, *American Technological Sublime* (Cambridge, MA and London: MIT Press), 1994.

[108] Maria B. Garda, Paweł Grabarczyk, "Technologiczna wzniosłość demosceny," in *Sztuka ma znaczenie*, eds. Dagmara Rode, Maciej Ożóg, and Marcin Składanek (Wydawnictwo Uniwersytetu Łódzkiego: Łódź, in print).

Fig. 2. Marcin Borkowski (in the foreground) during the Children's Day celebrations in 1975 at the Palace of Culture and Science in Warsaw

Source: Marcin Borkowski's private archive

Later, when Borkowski took the subject "Electronic Computational Techniques" while studying at the Faculty of Chemistry at the University of Warsaw, he submitted computational orders in the form of a properly punctured sets of perforated cards (one line of code per one card), which was done by throwing them into a box at the university. The results were ready by the next day. But the RIAD, which he actually used, was not only in a completely different building, but also in a different city district. The few who had direct contact with the computer were its full time operators. All respondents described contact with these large-scale calculating machines as requiring special care due to the devices' great sensitivity to even the slightest changes in the environment. Many respondents repeated anecdotes about operators "tiptoeing" around vulnerable RIADs so as not to interfere with the calculation process (e.g. by creating vibrations or changing the room temperature). In fact, Tomas (one of Kirkpatrick's Polish respondents) recalls that the mainframes "required the operators to wear special footwear."[109] This is confirmed by Borkowski, who corroborates that "the operator at Krakowskie [Przedmieście Street, in Warsaw], was wearing wooden clogs while in the computer chamber because he believed that better insulation from the floor made the electrostatics less problematic."[110]

---

[109] Kirkpatrick, "Meritums, Spectrums," 245.
[110] Private correspondence with Marcin Borkowski.

In the 1970s, most universities used calculating machines such as the already mentioned Odra or RIAD, or minicomputers such as MERA 300 or 400. However, by the 1980s, microcomputers were already being used. Waldemar Czajkowski studied electronics at the Wrocław University of Technology and remembers that there was talk among students that ZX-81 microcomputers would be used in class.[111] For most respondents born in the 1970s, their first contact with information technology was through a microcomputer. Those who were of school age during this decade most often had their first contact with a computer in a public place, often in a state institution such as a school, a community centre or their parents' workplace. Wojciech Nowak recalls that his first encounter with a microcomputer most probably took place on May 1, 1984 – Labour Day, which was proudly celebrated in the communist system. He went with his parents to visit the nearest production plant, Zakłady Azotowe Tarnów-Mościce, then bearing the name of Feliks Dzierżyński. Nowak, who was still a little boy at the time, recalls that among the large machines, which were likely used to automate production processes, there was also a microcomputer. He does not remember what model it was, but he emphasises that "it was a great experience, that I could see ... and even touch it."[112] The possibility of even a very limited interaction with a computer, where one could write something and see what they typed on the screen, aroused great enthusiasm among the young.

The impact of the state-owned enterprises on the diffusion of new media in the PRR was multifaceted. It was not limited only to the use of modern devices at work or in a professional context in the broader sense. In the socialist economy, local enterprises had social responsibilities and, among other activities, sponsored community centres where there was access to many forms of media culture, not just computers. For example, Grzegorz Juraszek remembers film screenings organised by the local housing association, which took place in the clubhouse "Cegiełka" ("The Brick") in Racibórz. During the summer, as part of an initiative for children spending their holidays in the city, "Cegiełka" hosted dedicated screenings of movies recorded on VHS (e.g. the Indiana Jones series).

Paweł Grabarczyk first encountered a computer around this time, but in different institutional circumstances, namely at his school. He went to Primary School No. 36 in the industrial city of Łódź, sometimes dubbed "Polish Manchester." The school building was erected on the 20th anniversary of the PPR and located amongst nineteenth century tenements in the old city centre. During a maths lesson a ZX Spectrum unexpectedly appeared in the room, which the teacher brought in from their home. After a short presentation, the pupils

---

[111] Waldemar Czajkowski, interview, May 3, 2016.
[112] Wojciech Nowak, interview, September 21, 2014.

could play a video game. The respondent still remembers playing an unofficial port of the arcade hit *Space Invaders* (Taito, 1978), and this experience contributed to the respondent's lifelong fascination with new technology. As he recalls, "I fell in love with it. I really knew: this is it!"[113] From that moment little Paweł was trying to get his hands on a machine of his own, constantly nagging his parents: "We must have it!" However, like many of his peers, he had to wait a long time before he could have a microcomputer at home. At that time in Poland, the purchase of such a device was associated with both financial sacrifices and surprisingly complex logistics, which we will discuss in the next section.

Another respondent from Łódź, Arkadiusz Staworzyński, was looking forward to having a personal device so much that he decided to build his own model. In one of the technical magazines from that period, he found a photo of the ZX Spectrum computer and drew it on a piece of paper using coloured felt-tip pens, with the intention of "practicing programming." Staworzyński says that it was a manifestation of his "extreme fascination" with the technology, although the motivation was strictly pragmatic. He thought that if he learned to "type" on this keyboard, and thus memorised the position of the keys, in the future it would be easier for him to do it in reality.[114] Coincidentally, in the same city, Paweł Grabarczyk independently came up with a very similar setup, as he recalls: "I was reading the listings and then *run* them on my cardboard machine trying to anticipate their output and draw it on the [imaginary] *screen*."

Around 1986, in response to the needs of local youth, the Wrocław Institute of Electronic Computing Technology (ZETO) created a microcomputer club called "Mikrozeto," where there was a charge for using the computers. An hour of ZX Spectrum use cost PLN 150, and an hour with Commodore 64 was PLN 200 (a then-equivalent of two issues of *Bajtek*). Game libraries were available on site, but as Artur Ciemięga recalls, the loading time of the programme was included in the time users paid for, which displeased the young clientele.[115] "Mikrozeto" was crowded and one had to reserve a seat even a week in advance. For Waldemar Doros, who didn't buy his own computer until the late 1980s, it was the site of his first direct contact with an 8-bit machine. As he recalls, "I was holding *Bajtek* in my hand and I was very excited." Waldemar could finally sit in front of the device, which previously he had only seen from a distance on television, as part of programmes like *Sonda*. Similar events were also organised by universities, and Doros participated in one called "Holidays with a computer" at the Wrocław University of Technology.

---

[113] Paweł Grabarczyk, interview, February 16, 2014.

[114] Arkadiusz Staworzyński, interview, November 22, 2014.

[115] The loading of a programme could take up to several minutes, and this process was prone to faults and often required repeated resumption.

Fig. 3. URNiK's interdisciplinary summer camp at the Garbaś Lake in 1987.
From the right: Piotr Górka, Barbara Jaroszewska, Józef Kapłanek (facing away)

Source: Andrzej Grossman's private archive

Another similar initiative was the Student Scientific and Cultural Movement (URNiK), which operated under the supervision of the Polish Scouting and Guiding Association (ZHP). Roland Wacławek was an active figure in this movement. Summer camps and winter holidays with a computer were organised as part of URNiK, during which young people could get familiar with the device or learn the basics of programming. Other organisations and institutions ran similar initiatives: Microcomputer Club Abakus, probably one of the first computer clubs of this kind, had been organising such trips since 1983.[116] URNiK's activities, however, seem particularly interesting because they were one of the elements of a wider IT education project, of which the magazine *Informik* was also a tool. Cooperation between the magazine and the scouting association resulted in the initiative having a very wide reach. For example, the Scouts Winter Computing Camp, an event run across January and February 1985 by the National Microcomputing Club INFORMIK (under the auspices of the *Młody Technik* journal), aroused such interest that "the number of applications was several times higher than the number of places."[117]

---

[116] Jarosław Kaczyński, "Mikrokomputer w lesie," *Przegląd Techniczny*, no. 42 (1985): 26.

[117] Grzegorz Zalot, "Kluby mikrokomputerowe: 'Informik.'" *Przegląd Techniczny*, no. 10 (1985): 28.

How were microcomputers obtained?

As I wanted to illustrate in the previous section, before the introduction of microcomputers, contact with information technology was an experience for the chosen few. Even for them it was often limited to being a spectator, deprived of the possibility of interaction. Artur Ciemięga recalls how he was looking at an Atari computer displayed in the window of a Pewex shop, where the device was even more expensive than in the Składnica Harcerska (Scout Depot).[118] As he says, it was "a coveted device, but no one could actually afford it."[119] Indeed, as Wacławek asserts, in those days people used to joke that a *citizen* is someone "who constantly has to *do without* something."[120] In Polish, the wordplay here is based on similarly sounding words: "citizen" (<u>obywa</u>tel) and "to do without" (<u>obywa</u>ć się), which provokes a reinterpretation of the term "citizen" as a – literally speaking – "do-withouter."[121]

Fig. 4. Queue of people willing to buy Timex computers in the scout depot near the editorial office of *Świat Młodych* in Warsaw, around 10th December 1986

Source: tok, "Komputer nie śledź," *Świat Młodych*, December 18, 1986, 1

---

[118] Składnica Harcerska (Scout Depot) was a national chain of stores specialising in scouting supplies but also catering to DIY enthusiasts.

[119] Artur Ciemięga, interview, December 4, 2015.

[120] Roland Wacławek, interview.

[121] Paweł Grabarczyk has suggested a more metaphorical translation, in which we refer to a "citiZEN" as someone who has to employ a constant zen-like perspective due to the lack of everyday goods. (Personal conversation, May 14, 2020.)

It seems that the shortage economy also permeated everyday language, and respondents used terms such as "they threw TV sets" [for sale][122] or "to hunt down a computer." Given the economic situation of the PPR, the main challenge faced by ordinary citizens wanting to buy a microcomputer was needing to gather appropriate resources. Roland Wacławek wrote in 1983's *Młody Technik* that the "domestication" of computers is only a matter of price."[123]

In the interviews conducted for this chapter, stories of journeys undertaken to buy microcomputers came up repeatedly. Often people travelled to a Pewex store, computer fair or a bazaar in a larger city. As Halina Bednarska recalls, referring to the last decade of the PPR, "back then nothing was simply bought, people had to go to great lengths to get anything."[124] In the PPR, "[the] retail sector was characterised by large, cooperative and state-owned trading conglomerates even in the 1980s [...] and riddled with endemic shortages, lines of customers and poor assortment."[125] The experiences previously shared by the future owners of TV sets, tape recorders or everyday food products such as sugar, in the second half of the 1980s also became the experience of those able to buy a microcomputer. As the editors of *Świat Młodych* wrote, there were often long queues in front of the stores surrounding the editorial office, but most often for the fish monger, rather than the scout depot. As they report, "if, running to the editorial office early in the morning, we encounter an obstacle in the form of a queue blocking the entire pavement, we think: aha, herrings, cod or something else from the distant seas."[126] This time, however, the reason for the queue was a delivery from Portugal – the Timex microcomputers that were compatible with the ZX Spectrum. As the editors continue:

> people have been crowding since dawn. It is hard to say whether to buy or to watch what's going on. Probably not only to watch, because since the shop opened its doors, a number of units have been sold. For now, however, we invite teenage fans… to… explore the store, since probably neither them nor their parents can afford such a purchase.[127]

---

[122]   This saying pertains to the stocking of limited quantity and thus highly desirable merchandise.

[123]   Roland Wacławek, "Mikrokomputery w natarciu," *Młody Technik*, no. 7 (1983): 17.

[124]   Halina Bednarska, interview, January 23, 2017.

[125]   Nebahat Tokatli, "A comparative report on the profiles of retailing in the emerging markets of Europe: Turkey, Poland, Hungary, Portugal, and Greece," *Journal of Euromarketing* 8, no. 4 (2000): 81.

[126]   tok, "Komputer nie śledź," *Świat Młodych*, no. 151 (1986): 1.

[127]   Ibid.

At a rough estimate, the price of a computer was as high as an annual salary, although the value of specific devices decreased over time. To buy a ZX Spectrum, Zbigniew Rudnicki had to sell his car, one of the aforementioned prides of the domestic automotive industry, Syrenka.[128] Only five years later Waldemar Doros was able to save enough money to buy his dream machine, the "Spektruś" as he called it (the English equivalent would be a "Speccy"), after taking up a summer job. It is also worth remembering that the computer itself was only one of the elements of the computer set. One also needed an output device, initially a TV and later on, especially with 16-bit computers, a screen. In addition, a tape recorder was needed as a data carrier, and ideally a joystick for input. Paweł Sikorski recalls that he was scolded by his parents because he did not inform them that in order to function properly, the microcomputer also needed the peripherals listed above, and that was an extra cost.[129]

Previously, computers could only be bought as part of consumer tourism, but since the mid-1980s they could be purchased in many places, although not in regular electronics stores. The most official places they could be purchased from were "hard currency" stores: Pewex, Baltona (for sailors) and Carbon (for miners). In addition, computers were available in consignment shops or BOM-IS[130] stores selling defective equipment, as well as in the aforementioned Scout Depot. Finally, for the best prices, one could buy computers from bazaars, such as the famous Jarmark Perski or the first computer fair at Grzybowska Street in Warsaw. The prices there used to fall in October when people were coming back from holidays and selling equipment purchased abroad.[131]

According to Kluska, the main ways to obtain a microcomputer before 1985, was to "have family living in the West, business trips, and procuring them from sailors coming back from overseas."[132] Buying a PC clone from Taiwan or ordering one in the Federal Republic of Germany was an option open only to holders of foreign exchange accounts[133] and was usually organised by the Polish-foreign joint ventures.[134] A more common practice was to travel to West Berlin, as Wasiak

---

[128] Zbigniew Rudnicki, interview, November 23, 2015.

[129] Paweł Sikorski, interview, November 28, 2016.

[130] BOMIS, or Biuro Obrotu Maszynami i Surowcami (Bureau of Machine and Resource Circulation), was a state enterprise facilitating the redistribution of surplus stock, production waste and defective goods to other state enterprises (since the 1950s) and private customers (since the 1970s).

[131] See Grzegorz Majczak, "Nowości z mikroświata," *TOP*, October 16, 1987: 16.

[132] Kluska, "Komputeryzacja jakby od końca," 387.

[133] See Jerzy Klawiński, "Jeden pasterz," *Informik*, no. 1 (1987): 4.

[134] In 1979, the PPR's authorities sanctioned the Polish-foreign joint-venture (the so-called "firma polonijna") as a new type of business allowed within the otherwise

points out, "[i]n the 1980s, paradoxically, Poles could travel there much more easily than GDR citizens."[135] Thus, not surprisingly microcomputer prices on the other side of the Berlin Wall were regularly published in the *Młody Technik*. A story of purchasing a Commodore 64 behind the Berlin Wall is even evoked in one of the interviews with the leader of the popular Polish music band Kombi.[136] Sławomir Łosowski used the microcomputer along with a Musical Instrument Digital Interface (MIDI) extension for sound sequencing.

As Tomasz Sielicki reports:

> in the early 1980s, many Polish IT specialists started making money by importing computers. [...] At the beginning, importing one computer was enough to buy an apartment.[137] In the years 1985–1987, such an operation would suffice for a car.[138]

From an interview conducted by the Save the Floppy team, we can learn that Ryszard Kajkowski, who became a pioneer of the Polish private IT sector while doing his PhD at the University of Gdańsk, would go for seasonal jobs to West Germany, already in the late 1970s.[139] As he recalls, in Hanover, he established a close connection with the Apple User Group Europe (AUGE). In 1980, thanks to the AUGE networks, he was able to buy an Apple II machine and then managed to successfully smuggle it into Poland. Using his in-

---

very limited private sector. "After the year 1980 Polish-foreign joint ventures became an important element of the Polish economic landscape. They were enclaves that worked for profit, and they were at least partly exempt from the central planning policies... In the late 1980s there were more than 700 such companies, and they employed about 100.000 people." Maciej Bałtowski, Szymon Żminda, "Sektor nowych prywatnych przedsiębiorstw w gospodarce polskiej – jego geneza i struktura," *Annales Universitatis Mariae Curie-Skłodowska. Sectio H, Oeconomia* 39, no. 4 (2005): 56–57.

[135] Patryk Wasiak, *Playing and copying*, 132.

[136] RetroKomp, "Rozmowa z założycielem KOMBI Sławomirem Łosowskim po koncercie na RetroKomp 2016," video, 3:28, accessed January 30, 2020, https://www.youtube.com/watch?v=jsEWsJull5k

[137] Of course we are not talking here about a ZX Spectrum but a high-end PC clone, which, sold through the BOMIS chain, would make the transaction much more profitable.

[138] Krystyna Karwicka, "Artyści i rzemieślnicy," interview with Tomasz Sielicki, November 18, 1991, accessed January 30, 2020, http://www.computerworld.pl/news/315398/Artysci.i.rzemieslnicy.html

[139] Tomasz Jachnicki, and Sebastian Stecewicz, "'Komputeryzujmy się.' Wywiad z Ryszardem Kajkowskim," June 13, 2016, accessed May 19, 2020. https://savethefloppy.com/2016/06/13/komputeryzujmy-sie-wywiad-z-ryszardem-kajkowskim-czesc-i.html

ternational connections, he became a subcontractor for a Japanese company and started writing commercial software. In 1980, in order to legalise his endeavour and register his future company - Computer Studio Kajkowski (CSK) – he had to join the local tradesmen association, the guild of master craftsmen in Gdańsk. As he remembers, during the interview he was asked how many journeymen he would employ... [140] Zbigniew Rudnicki had a very similar experience while registering his software company in Katowice in 1986, also through the local tradesmen association. [141]

For IT professionals, the high price of microcomputers was a challenge but also an opportunity. For regular consumers it meant that the microcomputer was a luxury good, and thus became a status symbol. As Halina Drachal wrote in *Głos Nauczyciela*, owning such a device was "in vogue." In her opinion, the computer somehow "ennobles, making a person feel like the chosen one, someone better, as if, in old-fashioned terms, they are a member of the upper classes." [142] Thus, being the first person in a group of friends to have a computer was associated with social capital. As Tomasz Miądzik recalls, "the person who had a PC had the most friends." [143] Wasiak observes that computers were also appealing as "Western material artifacts," and all products "manufactured in the West were regarded as superior to domestic products, and their possession and consumption was recognised as a significant indicator of both material and cultural capital." [144]

However, as Wacławek notes years later, the microcomputer "ennobled" its owner on completely different terms than a prestigious car, because driving a car was much easier to master than learning how to use a computer. [145] Another respondent recalls the story of friends who bought a microcomputer at Pewex, but soon decided to sell it because they did not know how to use it. [146] Low consumer awareness was also widely commented on in the press. In an article on the purchase of a personal computer, Grzegorz Majczak sums up that "most owners of these supersets have no idea what a ZX Spectrum could be used for (except games), let alone an IBM AT or Toshiba 3100." [147]

---

[140] Ibid.
[141] Zbigniew Rudnicki, interview.
[142] Halina Drachal, "Flirt z komputerem," *Głos Nauczycielski* no. 1 (1987): 7.
[143] Tomasz Międzik, interview, November 8, 2015.
[144] Wasiak, "Playing and Copying," 131.
[145] Roland Wacławek, interview.
[146] Tomasz Grochowski, interview, November 15, 2015.
[147] Majczak, "Nowości z mikroświata."

Computers, counterintelligence and secret police

The Security Service also became interested in the influx of computers from abroad. From 1987–88, the Provincial Office of the Interior in Poznań conducted an investigation into people who brought computers to the PPR.[148] The Head of the Department of the Second Provincial Office of the Interior (the department responsible for counterintelligence), assessed the situation in a letter from June 24, 1987:

> Our intel clearly shows that Western intelligence services equip their agents with new means of communication – personal computers. These devices are used to conceal, store and transfer information to the Foreign Intelligence Headquarters in a strictly coordinated manner. Due to the mass influx of computers to Poland, there is certainty that foreign intelligence uses this natural opportunity to equip their agents operating in the country. Therefore, there is a need to deal with this problem...[149]

The officers conducting the investigation interviewed people who, when returning home from abroad, brought a microcomputer to Poland. One of the interviewees reported that during a tourist cruise on a ship belonging to the Polish Ocean Lines, while stopping over in Brussels, he bought a ZX Spectrum Plus computer for around USD 100. Interestingly, five of his six fellow travellers also purchased computers. The officer noted that the computer acquired by the interlocutor "is used only for his own purposes, as a toy."[150] We can suspect that the main reason of the computer purchase was to play video games.

Based on mine and Kluska's research in the archives of the Institute of National Remembrance, it seems that the SB was not interested in microcomputers as a tool for the reception and production of cultural forms, such as video games. The operations undertaken by the Security Service mainly concerned any mismanagement associated with the purchase of equipment or their use in workplaces that was contrary to their intended purpose.

Kluska describes the so-called "Afera komputerowa" ("Computergate") as probably one of the largest operations of such kind, uncovering fraud on a national scale and involving international participation. As he writes, microcomputer speculators:

---

[148] IPN Po 06/281/27.
[149] IPN Po 06/281/27 362/17, p. 1.
[150] Ibid., 22.

imported computer hardware using various available routes, and then sold them to state-owned consignment stores, e.g. BOMIS or the DOMAR chains. Then the microcomputers would be bought at a much higher price by state institutions, which did not have foreign currencies at their disposal but needed an official bill of sale, not possible to obtain while, for example, buying at a bazaar.[151]

The whole operation would be coordinated, including hiring middlemen and bribing state officials, so the state institutions would buy the imported products. As Kluska reports, "Computergate" stemmed from a search of the student dormitory "Babylon" in Cracow, during which "shocked militiamen found 5M PLN, 8,000 USD and 4,000 DM and an Atari computer, under the bed of a third year Electronics student."[152] Soon the investigation tracked down other students involved in the scheme, citizens of Bolivia, Cuba, Panama, Yemen and Argentina. However, as Kluska observes, what is probably most shocking about this case is that the press covering "Computergate" expressed their support for the perpetrators.[153] After all, even though the students broke many laws, "national enterprises and institutions were equipped with – otherwise unattainable – modern computers (including the much in demand IBM machines)."[154]

Kluska also reveals the details of the SB action against microcomputer speculators at the University of Łódź (operation codename: "Bit").[155] The purchase of devices with relatively low computing power and small capabilities raised suspicion of recreational rather than professional use. Deployed operatives discovered that the most popular microcomputer among scholars seemed to be the ZX Spectrum, a machine mostly associated with gaming, and according to SB experts not suitable for higher education purposes. As Kluska concludes, "soon chairs of many institutes and departments at the tertiary education institutions of Łódź had to justify their peculiar hardware preferences."[156]

I have followed the files of a similar operation (codename "Takson"), conducted in relation to computers bought by the Technological University of Szczecin.[157] The SB files state that "[a]s a token of gratitude and a form of payment for conducted research, Mr Fujimoto, in 1986, sent an IBM computer, splitting it into 17 parts, dispatched to different recipients."[158] Based on the SB operative's

---

[151] Kluska, "Komputeryzacja jakby od końca," 390.
[152] Ibid.
[153] Ibid., 391.
[154] Ibid., 391–392.
[155] See ibid., 389–392.
[156] Ibid., 395.
[157] See IPN_Sz_00_11_1711.
[158] Ibid., 14.

findings, Mr Fujimoto bought the newest IBM PC clone from Hong Kong (in the files the machine is labelled as: "ARC PC XT") and then prepared it for a long freight to Poland on the cargo ship *Jacek Malczewski*, heading from Yokohama to Gdańsk (the journey took approximately 6 to 8 weeks). Splitting the computer into many parts apparently helped to avoid taxes. When sold at BOMIS at exaggerated prices, individual parts of the computer would not exceed 200,000 PLN in worth (the wealth tax threshold). Based on the SB documents we can suspect that the reassembled computer could have reached the final price of approximately 15M PLN. Furthermore, the investigation uncovered that Mr Fujimoto was actually a Polish-born Japanese citizen and a graduate of the Technological University of Szczecin. In the 1980s, while living in Tokyo, he started a successful business in the field of digital imaging. Based on close reading of his letters preserved by the SB, we can establish he was most likely trying to support his former Polish colleagues: financially and otherwise (e.g. by sending toys to a local kindergarten or sharing the newest scientific publications).

Although computers could pose a threat to the communist authorities as a tool used by foreign services, or an object of financial speculations and potential tax evasion, SB seemed to overlook a much more likely threat – the flow of cultural content from the West. There was also no evidence that the authorities surveilled microcomputer fans. Perhaps it was not noticed that several hundred thousand microcomputer users co-created a kind of cultural formation, or this formation was not perceived as ideologically harmful, just like other hobby or fan movements.[159]

Hobbyist movement

The hobbyist-driven appropriation of microcomputers in the 1980s is reminiscent of earlier amateur movements in pre-war Poland, such as radio amateurs,[160] as well as photographers. Each time, a grassroots culture of innovators and early adopters would start converging around topical magazines and literature, creating a secondary wave of diffusion of innovation. In the case of photography, these publications were mainly manuals and textbooks,[161] and in the case of radio, also

---

[159] Stanisław Krawczyk, "Gust i prestiż. O tworzeniu pola prozy fantastycznej w Polsce" (PhD diss., Uniwersytet Warszawski, 2019), 220–223.

[160] See Matthias Barelkowski, "Hobby bez granic? Rzecz o krótkofalarstwie w Polsce w latach 1925–1990," *Przegląd Historyczny* 109, no. 4 (2018).

[161] See Jan Rajmund Paśko, "Fotograficzne wydawnictwa seryjne a ruch amatorski w XX wieku," *Annales Universitatis Paedagogicae Cracoviensis. Studia de Cultura* 3, no. 112 (2014).

magazines such as the pre-war *Radioamator* (1924–27) and its post-war continuations. In some respects, however, the situation with microcomputers in the 1980s was unique because of the technological lag. For example, amateur photography at the turn of the twentieth century was an international movement with a similar pace of development around the world. Polish inventors of this period, such as Jan Szczepanik and Kazimierz Prószyński, created new technical solutions that were adopted and used by the institutionalising global industries of photography, film, and, in the long term, television. Unfortunately, the microcomputer hobbyists of the PPR did not get the chance to influence the international arena, but that does not mean that we were dealing with a movement that was not innovative or creative. From a cultural studies perspective, the most interesting aspects of the hobbyist movement were the various local cultural practices, as well as organisational and discursive structures that determined the specificity of the cultural appropriation of microcomputing in the late PPR. Among them are all the ingenious ways of bringing these devices into the country and their further distribution, equipment modifications dictated by the reality of the shortage economy, as well as fan communities developing around them, which in the future were to contribute to the development of the Polish demoscene, video games culture, and finally to the broadly defined creative industries based on digital media.

Tinkering out of necessity

8-bit microcomputers were devices that constantly reminded users of their materiality. They required a lot of attention as well as interim repairs. Less burdensome practices that were mentioned by the respondents include manual cranking of the audio tape (data carrier) with a pencil, or aligning the tape head with a screwdriver in order to read the game cassettes recorded by the given distributor. This was an element of the competition between local software distributors, who, for example, sold games at the Grzybowska computer fair. During copying, these distributors set the bevel of the head in a certain way as only duplicating their system guaranteed that the saved file could be read. This prompted many people to use the same distributor for each purchase, because then it was not necessary to adjust the head each time a new cassette was loaded.

Joysticks were even more troublesome because they often wore out quickly. Several respondents had the experience of building their own joystick from recycled materials, because buying another unreliable controller out of their pocket money did not seem like a good investment. Moreover, the presence of schematic joystick designs in the press materials of that period[162] indicates that this was not

---

[162] See for example Roman Poznański, "Drążek sterowy," *Bajtek*, no. 1 (1985): 27–29.

an isolated experience. For example, to build his control stick, Tomasz Cieśle-wicz used reclaimed materials including a flexible hydrogen peroxide bottle, parts of a disassembled toy (which served as contacts), and a wooden stick.[163] With this in mind, it is not surprising that in 1985, Krystyna Karwicka[164] wrote that Poles were "hobbyists out of necessity." Karwicka pointed out the very high cost of services in the PPR, and speculated that "we will necessarily become a nation whose citizens will have to be specialists in many areas."[165] In fact, in the 1980s, in the face of constant shortages, hobbyist movements offered a relief to the crippled state machine. For that reason, even though not ideologically flawless, hobbies were usually sanctioned by the communist authorities, as it was in the case of allotment gardens. Until the 1950s, in the USSR, "individual gardening was seen as the legacy of so-called *petit-bourgeois* lifestyle that had to be eliminated,"[166] yet the harsh realities made the Party change its stance. In the PPR, especially after the introduction of martial law, when the nation "was reeling from food shortages, [...] [the allotment gardens movement] benefited the state by shifting some of its direct responsibility for the public's food security to a non-governmental group."[167] The situation was similar in other areas of the economy.

Building your own microcomputer

Before he got involved with *Młody Technik* magazine and the popularisation of microcomputing, Roland Wacławek belonged to another hobbyist movement. As a teenager growing up in the mining district of Silesia, he was an avid radio amateur. He regularly read technical magazines (e.g. *Horyzonty Techniki*, *Radioamator* or *Radioelektronik*, as well as Soviet *Radio*), and constructed his own electronic devices, including radio receivers and various pieces of equipment for measuring and use in workshops (e.g. power supply units, generators, oscilloscopes, wobbulators, voltmeters and frequency meters). He also built electroacoustic equipment, and even a fully functional television set. In the early 1980s he decided to take on a completely new challenge. During compulsory military service, when he was stationed as an officer cadet in the air force technical services, he decided

---

[163]   Tomasz Cieślewicz, interview.

[164]   Krystyna Karwicka-Rychlewicz (1934–2016) was a journalist and social activist, since the 1970s professionally associated with the technical and, later, IT press.

[165]   Krystyna Karwicka, "Hobbyści z konieczności," *Przegląd Techniczny*, no. 41 (1985): 31.

[166]   Louiza M. Boukharaeva, Marcel Marloie, *Family Urban Agriculture in Russia: Lessons and Prospects* (Cham: Springer, 2015), 150–151.

[167]   Anne C. Bellows, "One Hundred Years of Allotment Gardens in Poland," *Food and Foodways* 12, no. 4 (2004): 262.

to construct an 8-bit microcomputer for personal use in his spare time. Earlier, during his studies at the Silesian University of Technology, he encountered the Polish 8-bit computer MOMIK 8b (MERA 1974). However, this was based on a transistor design (TTL) and ferrite memory – not microchips. Nevertheless, thanks to his acquaintances at that time, Wacławek could obtain more advanced components, including processors and RAM memory imported (not always legally) from the Soviet Union. In the evenings from the army barracks, he worked on the construction scheme of his future microcomputer, and during Saturday and Sunday home leave he meticulously designed circuit boards.

Fig. 5. Ilona Wacławek using the microcomputer WAC-80, constructed in 1981

Source: *Młody Technik*, no. 12 (1984): 30

Wacławek's microcomputer (named WAC-80) was initially based on a Soviet-made microprocessor (equivalent to INTEL 8080), and later, when an upgrade was possible, a Z80 class processor. Other elements were obtained thanks to his extensive social network of hobbyists interested in electronics. Wacławek spent many months hunting for parts, regularly visiting the electronics market located at what was known as Dom Technika[168] in Katowice. The weekly market was

---

[168] Dom Technika was a common name given during the communist era to the regional headquarters of what was then the Central Technical Organisation (Naczelna Organizacja Techniczna – NOT), and what today holds the name of the Polish Federation

organised in collaboration with BOMIS and regular customers could get bargain price electronics which had failed quality controls but were often at least partially functional. Wacławek's experience suggests that the majority of the available products were fully functional, only requiring initial testing.

After over a year of preparations, and right in time for Christmas 1981, a fully assembled microcomputer WAC-80 awaited its designer at home, sitting in a neatly fitted wooden case. However, on December 13, just a week before the end of Wacławek's military service, the infamous martial law was declared. It was the authoritarian government's final attempt to choke the political opposition (i.e. the Solidarity movement), and would last until 1983. Wacławek was ordered to remain in service for another hundred days. As he recalls, he got a notebook and used this time to draft software for his device.

Wacławek would go on to upgrade his machine several times. For example, he managed to get a scrapped disk storage unit which after a thorough conversion was connected to the computer, offering a "dizzying" capacity of 2MB. As he recalls, the most challenging technical issue of WAC-80 proved to be the treatment of the hydraulic head drive. Later, in 1983, the machine was featured in an issue of *Młody Technik* as an encouragement for other hobbyists to build their own microcomputing machines. However, only one unit of WAC-80 was ever constructed and the machine has not survived to the present day. It was damaged beyond repair in a basement flood at Wacławek's house.

It is worth noting that during the conceptual work process on WAC-80, Wacławek used materials available in Russian, which were actually exact translations of Western publications. It would seem that knowledge of English was not crucial for hobbyists interested in new technologies. Indeed, Wacławek's example illustrates that the patterns of knowledge circulation behind the Iron Curtain were often not obvious or straightforward. In the 1980s it would be very hard to get the latest American publications on microcomputing, as doing so would require having a social connection in an Anglo-Saxon country who was able to send the materials to Poland. However, thanks to the Cold War technological race, Anglo-Saxon publications were brought to Moscow relatively quickly. Following this, Russian reprints were developed unscrupulously and at an express pace, and they could be purchased at Polish bookstores shortly afterwards, often

---

of Engineering Association. During the PPR, the association would welcome professional engineers and technicians with a STEM degree (secondary or tertiary), and it would support the development of several new higher education institutions. The organisation has a long tradition, dating back to the Polish Polytechnical Society, founded in 1835 in Paris by Polish emigrants and political exiles, many of whom left Poland after the November Uprising (1830–1831). See https://not.org.pl/o-not/historia, accessed May 19, 2020.

in stores that specialised in foreign literature throughout Katowice. As Wacławek recalls, he would also seek Soviet technical magazines, as they were worthy of merit too (e.g. *Radio* or *Modelist-konstruktor*).

There were certainly more hand-made microcomputers in the early 1980s and more builders like Wacławek in the PPR. Kluska mentions at least some of them, such as a ZX80 based OPOL-1 (1984) made by three students at the electrician trade school in Opole (Jacek Domerecki, Marek Krokoszyński and Aleksander Sachanbiński), or a ZX Spectrum compatible Kubuś built by a graduate of the AGH (University of Science and Technology) in Cracow.[169] Particularly noteworthy in this context is the COBRA-1 project (modelled on the ZX-81) by Andrzej Sirko, a diagram of which was published in installments in the magazine *Audio-Video* throughout 1984. Although assembling this set required advanced technical knowledge and obtaining rare components, the undoubtable advantage was the price, because "it required the equivalent of two average salaries, while the cheapest second-hand [ZX] 81 cost three times as much."[170] It is difficult to say how many COBRA-1 computers were created in the 1980s, although it was probably not a significant number. Some of them have survived. For example, one unit is in the collection of the Museum of Computers and Information Technology in Katowice. Other stories that are yet to be explored can be found on social media (e.g. the "Polskie komputery" project[171]).

From ZX Spectrum to "the last bastion of Atari"

In the first phase of 8-bit microcomputers diffusion after the CoCom embargo was withdrawn, people interested in buying them did not ask themselves what model to choose, but rather how to get a computer at all. Grabarczyk recalls that he was one step away from buying a Meritum computer, which he did not want. However, after months of prolonged unsuccessful efforts to obtain a microcomputer, he was almost willing to accept such a solution. Why was the Meritum not very attractive to 10-year-old Paweł? As he explains, "I would have a machine that none of my friends had, the software was very hard to get hold of and who knows if I would not become discouraged [towards computers]."[172]

The beginnings of the computer market in Poland are primarily connected with the ZX Spectrum, and that was essentially determined by price

---

[169] Bartłomiej Kluska, *Automaty liczą*, 157–160. See also Michał Liszka, "Opol-1 i co dalej," *Komputer*, no. 1 (1986): 40.

[170] Kluska, *Automaty liczą*, 159.

[171] "Polskie Komputery," accessed January 30, 2020, https://www.facebook.com/polskiekomputery/

[172] Paweł Grabarczyk, interview.

and availability – in the shortage economy, other factors were of marginal importance. Most respondents also mentioned ZX Spectrum as their first "machine." However, another platform, Atari, soon took over. Lucjan Wencel,[173] a Polish immigrant living in California, began to import Atari microcomputers to Poland thanks to the professional cooperation of Jack Tramiel,[174] who was then the head of Atari.[175] Initially he imported the Atari 800XL, which could be bought at Pewex for USD 120, but later he switched to the Atari 65XE. Atari was the only Western company to offer a licensed service in Poland, and service points (under the aegis of PZ Karen) were located in Warsaw and Cracow. The availability at Pewex and the possibility of warranty repair determined the dominance of this platform on the market.

Atari's domination in Poland, like the primacy of other platforms in other countries of the Eastern Bloc (e.g. ZX Spectrum in former Czechoslovakia or Commodore 64 in Hungary), was in many respects the result of a coincidence, as was often the case when one platform got to the market first, or had the most widely available software. In an interview given recently, Lucjan Wencel simply said: "I was aware that I would sell as many of these computers as there would be software for them. As such, I procured games for Polish sellers and various computer studios to copy and distribute them at bazaars and computer fairs."[176] This information is very interesting, because it not only allows us to guess how software made its way to the Polish market (also for other platforms), but also indicates a business strategy that is somehow the opposite of the model used by the producers of game consoles: give equipment, earn on software.

Interestingly, in the last decade of the twentieth century, Poland was what Kluska calls: "the last bastion of Atari."[177] In fact, the last games released on the 8-bit Atari in Poland, such as the cassette edition of *Tekblast* (Sikor Soft, 1998), were probably the last productions released on this platform in the first

---

[173]   Lucjan Wencel, born in 1949 in Toronto, a physicist best known as the founder of the Polish community company PZ Karen and two American companies: California Access and California Dreams. These entities were key to the computerisation of the PPR and to the development of the local computer games market.

[174]   Jack Tramiel, born in 1928 in Łódź was an American entrepreneur of Polish-Jewish origin, founder of Commodore and later director of Atari. Born Idek Trzmiel, he was a survivour of the Łódź ghetto and forced labour camp in Ahlem. He died in California in 2012.

[175]   Piotr Mańkowski, *Cyfrowe marzenia. Historia gier komputerowych i wideo* (Warszawa: Wydawnictwo Trio – Collegium Civitas, 2010), 172.

[176]   Piotr Mańkowski, "Polak z Doliny Krzemowej. Wywiad z Lucjanem Wenclem," *Pixel*, no. 1/55 (2020): 49.

[177]   See Kluska, Rozwadowski, *Bajty polskie*, 111–126.

period of its global commercial operation.[178] This does not mean that Poland was the laggiest country in the world when it comes to the rate of computerisation. The extraordinary vitality of the Polish Atari scene was closely related to the activity of the local Atari demoscene groups. Next to the ZX Spectrum and 8-bit Atari, the Commodore 64 was a very popular platform. There were also other brands on the market, as numerous companies run by Polish-foreign joint-ventures manufactured their own constructions based on Western models. For example, Lucjan Wencel's PZ Karen released their own PC clones under the name Quasar, and Ryszard Kajkowski's CSK produced Lidia computers compatible with Apple II. Finally, at the bazaars and computer fairs, customers could find "various compilations produced by amateur DIY enthusiasts from the so-called 'Taiwanese imports' [IBM PC compatible custom builds] and valued at a guess."[179]

Generational gap

As Barbara Ostrowska recalls, regarding the use of the computer at home, "parents were afraid that something would break."[180] There was indeed a generational gap in how new technologies were perceived and used. The editors of TOP wrote: "Joystick, interface, light pen, Basic – these are words that our children use with ease. For us adults, however, these sounds are foreign, sometimes even a little disturbing."[181] This does not mean, however, that adults were not interested in new technology. An anonymous reader of Przegląd Techniczny reported in a letter to the editor that "there are no educational programmes on TV for older generations who want to and can computerise the industry."[182] It seems that this reader would have liked to use a computer for professional purposes, in a way not common among the youth of the time.

Considering the average age in modern retrogaming communities, it can be concluded that the generation for whom microcomputers were an important life experience were people born in the 1970s, which is to say those who were children and teenagers in the 1980s. The indoctrination of youth in totalitarian and authoritarian countries has always been especially important to the regime. On the one hand, as Marek Wierzbicki says, the authorities of the PPR were "constantly striving to gain influence on the young generation and its way of thinking,

---

[178] See Maria B. Garda, Paweł Grabarczyk, "'The Last Cassette' and the Local Chronology of 8-Bit Video Games in Poland," in Games and the Local, ed. Melanie Swalwell (in print).

[179] "Rynek komputerowy," TOP, February, 1987.

[180] Barbara Ostrowska, interview, November 28, 2015.

[181] JŁ, "Sargon II (recenzja gry)," TOP, December 18, 1987: 10.

[182] "Komputer dla przemysłu," Przegląd Techniczny, no. 47 (1986): 19.

to bring it in line with the authorities' expectations."[183] On the other hand, as Jerzy Eisler claims:

> ... in the 1980s this refractory or hostile youth was already a mass phenomenon, counted in millions. There were sometimes almost entire universities and high schools in which no regime youth organisations operated, or at most there was just a group of activists. You could not put them all in jail, ignore them, or just pretend they did not exist. It was also impossible to laugh at and ridicule the majority. What I mean is that the group of non-members was so large in the 1980s that the authorities did not know what to do with it. Hence my concept of 'skipping a generation' – the rulers' realisation that this 1980s generation of youth was simply lost to them. It seems to me that they understood that they had 2–3 million young people against them and that they had no chance of winning this war.[184]

On the one hand, the communist authorities were aware of the expectations of the Polish youth, as evidenced by a statement from the Minister for Youth Affairs, Aleksander Kwaśniewski,[185] in which he admits that he understands that the "youth dreams of the computer and modernity."[186] For example, significant funds were allocated to equip the camps for (ideologically) promising teenagers, called "the 21st century Avant-garde," with the latest computer equipment.[187] On the other hand, one should remember that the Party pursued its own goals. As Eisler observes, speaking of the PPR: "... youth organisations or institutions dealing with youth ... pursue the interests of those in power, not the youth."[188] Although there were ideas in the USSR to introduce home microcomputers modelled after Apple II, it turned out to be just wishful thinking.[189]

Journalists at *Przegląd Techniczny* warned in 1984 that "... without adequately preparing society, i.e. introduction of IT to schools, we will face

---

[183] Marek Wierzbicki, Jerzy Eisler, Joanna Sadowska, Łukasz Kamiński, "Młodzież w PRL. Dyskusja," *Pamięć i Sprawiedliwość* 10, no. 1 (2011): 16.

[184] Ibid., 17.

[185] Later, Aleksander Kwaśniewski became the President of Poland and served two terms (1995–2005).

[186] Grzegorz Onichimowski, Roman Poznański, "Generacja z komputera. Rozmowa z Aleksandrem Kwaśniewskim – ministrem d/s młodzieży," *Bajtek*, April, 1987: 3.

[187] Dariusz Magier, Michał Mroczek, "Program 'Awangarda XXI wieku' – w poszukiwaniu nowego modelu pasa transmisyjnego," in *To idzie młodość. Młodzież w ideologii i praktyce komunizmu*, ed. Dariusz Magier (Lublin and Radzyń Podlaski: Archiwum Państwowe, Towarzystwo Nauki i Kultury Libra, 2016).

[188] Wierzbicki, Eisler, Sadowska, Kamiński, "Młodzież w PRL," 20.

[189] See Zbigniew Stachniak, "Red Clones: The Soviet Computer Hobby Movement of the 1980s," *IEEE Annals of the History of Computing* 37, no. 1 (2015).

another barrier. And even if we can overcome the current problems with production, an unprepared society will not be able to use this equipment."[190] Nevertheless, as Roland Wacławek notices, however uncomfortable it was for the authorities, many pioneers of microcomputers, including the most famous ones who succeeded in the West, were self-taught. This claim is supported by the statements of respondents who learned to programme using textbooks and the collective intelligence of communities centred around computer fairs and microcomputer clubs.

Fig. 6. The famous computer club at the Palace of Culture and Science in Warsaw

Source: *Przegląd Techniczny*, no. 11 (1987): 43

Undoubtedly, some members of the ruling elite, especially the aforementioned Aleksander Kwaśniewski, noticed the growing potential of the emerging hobby movement. The establishment of the Microinformatics Clubs Council, which aimed to subject the fan movement to centralised supervision, proves this. However, these activities did not bring much success, because the formal

---

[190] Ewa Mańkiewicz-Cudny, Roman Dawidson, "Kto kocha komputery?," *Przegląd Techniczny*, no. 10 (1984): 18.

and informal clubs of microcomputer enthusiasts were so numerous and developed so spontaneously that it was impossible to effectively control them. From the point of view of the authorities, the clubs were intended to support the introduction of computers to schools, not to domesticate them. We must remember that the communist state wanted to impose role models on citizens not only in public, but also in private life.[191] However, it seems that such attempts in the late 1980s were increasingly difficult to implement.

Based on data provided by Wasiak, we can speculate that before 1989, there were more than 800 communist sanctioned computer clubs operating within state-sponsored cultural centres (not counting in schools). Although large in number, the clubs developed under the auspices of the communist state institutions did not survive the transformation period and soon "there was no mention of computers clubs in computer magazines."[192] What persisted were the user groups described by Wasiak, focused on specific computing platforms and often overlapping with the local demoscene, as well as gaming and hacking cultures.[193] These user groups would often meet at the computer fairs.

Computer fairs

In the second half of the 1980s, Polish society, and especially the youth, began to organise itself when it came to domestication of microcomputers. According to the often-quoted statement by Ryszard Pregiel, "next to the vegetable market, IT is slowly becoming the branch where market law begins to play its role."[194] It was part of the broader process of system collapse and the emergence of consumer and civil society. Analysing the reasons for this would require further sociological research, but it seems that young people were able to organise themselves in ways that the authorities could neither prohibit nor take control of. A good example of such a situation was the peculiar institution of computer fair (giełda komputerowa).[195] If we compare Polish computer fairs of the 1980s to the early Californian microcomputing scene of the 1970s, we will discover much more similarities to the Hombrew Computer Club then to the West Coast Compute Faire. The Polish fairs took place on a weekly basis and for regulars, it was

---

[191]   See Wierzbicki, Eisler, Sadowska, Kamiński, "Młodzież w PRL," 20.

[192]   Wasiak, "Playing and Copying," 139.

[193]   See ibid., 142–147.

[194]   Cited after Kluska, *Automaty liczą*, 195.

[195]   Computer bazaar is another term used in the literature. In my opinion, the term computer fair better encapsulates the wide variety of local hubs for computer hobbyist in Poland in the 1980s, some of which also shared characteristics with a baazar.

a unique possibility not only to exchange hardware and software, but also to discuss know-how, and simply build friendships and future collaborations.

The biggest and most influential computer fair was situated at the Grzybowska Street in Warsaw and established with the support of the *Bajtek* magazine. Wasiak reports that *Bajtek*'s editorial team viewed computer fairs "as sites where computer users, instead of entrepreneurs, should be able to sell their old computers and peripherals for fair prices," yet as he continues, "[i]n reality, peddlers moved in from other bazaars and quickly took over the supply of hardware and pirate software."[196] Grzybowska computer fair was located in the premises of the Szkoła Podstawowa Nr 25 im. Komisji Edukacji Narodowej (Commission of National Education Primary School No. 25). Inside of the building there was mostly software on offer, and outside, vendors would sell hardware, often directly out of their car boots. Undoubtedly, Grzybowska was the central hub of informal economies around computer use in Poland and provided supplies to other regional computer fairs. Paweł Sikorski recalls that for him, growing up in Warsaw, Grzybowska was actually the only place where one could get acquainted first-hand with technological innovations.[197] The Grzybowska computer fair was unique in many respects. I would argue, however, it should not be treated as a typical example of a Polish computer fair of the 1980s but rather a model of cultural practices that was later adopted to local needs. Depending on various local circumstances, regional computer fairs could share characteristics with a community club, a boot fair, an exchange or a bazaar.[198]

Although the peak of computer fairs was the early 1990s, already in the 1980s it was an important place for the development of (sub)cultures related to the use of computers, and thus gaming culture and the demoscene. However, it is worth remembering that the fairs were only located in the largest cities, and the inhabitants of smaller towns were dependant on much less institutionalised social networks related to their hobby. As Wasiak argues, "it can be assumed that many people in the provinces, despite their willingness and ability, did not learn programming or how to create graphics precisely because of the lack of social contact with people sharing their interests."[199]

Although this was in the period preceding the development of social media,[200] it is worth remembering how important the social aspect of owning microcomputers was. Owners of microcomputers with compatible software libraries would search

---

[196] Wasiak, "Playing and Copying," 133.
[197] Paweł Sikorski, interview.
[198] See Garda, Grabarczyk, "'The Last Cassette'."
[199] Wasiak, "'Grali i kopiowali'," 219.
[200] However, after 1989 BBS were already active in Poland. Earlier shows were incidental.

each other out, so they could exchange and copy programmes at events known as "copy parties." Frequently, these meetings were also where participants tested out new software, which most often meant playing games. As Wojciech Mikołajczyk recalls, "with no Internet, when people met up, they brought their computers with them."[201] The interviews repeat stories of such expeditions and contacts, including examples of international exchange. For example, Waldemar Czajkowski mentions that during his studies, as part of the Voluntary Labour Corps,[202] he partook in an exchange in Czechoslovakia and during the weekend he went to Ostrava to swap software with a person he had previously corresponded with.[203]

What were microcomputers used for?

According to Wasiak, "playing and copying games was definitely the most popular practice of Poland's personal computer users."[204] The interviews conducted for this project seem to confirm this. Paweł Grabarczyk recalls that, to him, the microcomputer seemed like a logical step in the domestication of arcade machines. Just as using VHS tapes and video recorders at home seemed a rational way of domesticating the cinema, so did the computer appear to be a good way to move the video arcades to one's own living room, he said.[205] From the standpoint of professionals, this mass consumer perspective was surprising, to say the least. Aleksy Kordiukiewicz said, "[f]rom the point of view of a Cinderella, we delight in children's microcomputers made by bankrupt companies and sold in the West for a couple of dollars."[206] And yet it was those "toys" that formed the first wave of the computerisation of Polish homes.

Kluska observed that "professional computer scientists, who were becoming increasingly oblivious to the needs of the steadily growing group of regular computer users, were on a different trajectory."[207] For those interested in professional computer science, microcomputers like the Commodore 64 or Atari 8-bit were just a child's toy, a "plaything" ("grywas"), as Piotr Fuglewicz, put it.[208] The computer game players from that period may have felt that this approach denigrated their

---

[201] Wojciech Mikołajczyk, interview, November 28, 2015.

[202] The Voluntary Labour Corps is a youth organisation focused on providing students with opportunities to acquire work experience; during PPR it was often relied upon by the state as a reserve workforce.

[203] Waldemar Czajkowski, interview.

[204] Wasiak, "'Grali i kopiowali,'" 209.

[205] Paweł Grabarczyk, interview.

[206] Aleksy Kordiukiewicz, "Cudowne mikrokomputery," *Przegląd Techniczny*, no. 1 (1987): 15.

[207] Kluska, *Automaty liczą*, 173.

[208] Piotr Fuglewicz, interview.

hobby. Still, if we look at the role 8-bit microcomputers play in publications about retrogaming platforms,[209] we may realise that the problem was not in the perception of microcomputers but in the games themselves. Older users treated computers seriously, as a tool for work, whereas young consumers associated them with entertainment. Video games became a new form of popular culture and various titles had the potential of creating what we would call today a fandom. In the West, media fandom started to emerge in the 1960s, introducing new cultural practices related to media consumption (e.g. Star Wars universe).[210] In Poland, the process accelerated in the 1990s, but was born in the last decades of the PPR. A good example of early media fandom convergence were the various fan convents and gatherings, organised for example by the Silesian Science Fiction Club, where the exchange of books and VHS flourished next to video games.[211]

It is interesting that very similar ideological and discursive tensions accompanied the domestication of computers in Australia. Basing her observations on an in-depth review of Australian computer and technology-themed journals, Melanie Swalwell remarks that "[u]sefulness was… equated with seriousness, and seriousness is about computing power and whether the computer was fit for the purpose for which it was envisaged."[212] She notes that the low price was suspicious to the public, which is obviously not as strong an argument in the Polish context, but which makes sense if we consider the prices of professional minicomputers. What is more, the limited power of the microcomputer caused doubts about its potential usefulness in the professional environment – that is, at work. In Swalwell's words, "[p]rogramming was the only use that was indigenous to the computer,"[213] which is why the hobbyists who produced new software, in fact, created new ways to use the computers.[214]

One of the most frequent uses mentioned by the producers of microcomputers was managing the household budget. However, both Wacławek and Swalwell

---

[209] Bill Loguidice, Matt Barton, *Vintage Game Consoles: An Inside Look at Apple, Atari, Commodore, Nintendo, and the Greatest Gaming Platforms of All Time* (New York: Focal Press, 2014), 105.

[210] See Francesca Coppa, "A Brief History of Media Fandom," in *Fan Fiction and Fan Communities in the Age of the Internet*, ed. Karen Hellekson and Kristina Busse (Jefferson, NC: McFarland, 2006).

[211] For more on relations of the early Polish fandom and microcomputing hobbyists, see Aleksandra Wierzchowska, "SF jest ulubioną rozrywką informatyków. Polski fandom a popularyzacja elektroniki," in *High-tech za żelazną kurtyną*.

[212] Melanie Swalwell, "Questions about the Usefulness of Microcomputers in 1980s Australia," *Media International Australia*, no. 143 (2012): 66.

[213] Ibid., 69.

[214] Ibid., 69–70.

point out that this was a "boring" and "cumbersome" activity, hardly popular among the users. Roland Wacławek lists, among others, the following uses of the microcomputer, which could be of interest to amateurs coming from other hobbyist movements: controlling the motions of a miniature train (modelling), "automating the giving and receiving of Morse signals" (two-way radios) and "calculating exposition times" (amateur photography).[215] In his 1987 book, from which the motto for this chapter was taken, he also made the following, highly symptomatic claim:

> Writing computer game software is not at all the exclusive domain of specialists. It is precisely in games that inventiveness and imagination matter the most. What is not needed is comprehensive knowledge in higher mathematics or economics, as is the case with many of the more serious programmes.[216]

In many ways, the roots of the Polish game industry can be traced back to communities of knowledge and user groups founded in the PPR around computer fairs.[217]

If we look at the global uses of microcomputers – not just in Poland – from our current perspective, we will see that their actual main uses were the cultural practices related to the production, distribution, and use of computer games. Furthermore, even now there are games still being made for many platforms from the past, attracting older users (whose experiences are similar to those of our respondents) as well as entirely new enthusiasts. Therefore, one can assume that the cultural appropriation of those microcomputers – which are now nearly 40 years old – is still in place and will remain so in the future.

Conclusion

As late as 1989, *TOP* magazine reported that:

> [i]t is not easy to purchase a computer, disk drive or printer in Poland. There are no Western shops here where one might freely choose between Atari, Commodore, and Amstrad. At the same time, the number of computers in the homes of people who enjoy buying these things indicates that such purchases are made eagerly and often.[218]

---

[215] Wacławek, "Z mikrokomputerem…," 129–135.

[216] Ibid., 145.

[217] For more on the origins of Polish video game industry, see Piotr Marecki, Tomasz "Tdc" Cieślewicz, *Oni migają tymi kolorami w sposób profesjonalny: narodziny gamedevu z ducha demosceny w Polsce* (Kraków: Korporacja Ha!art, 2020); Marcin Kosman, *Nie tylko Wiedźmin. Historia polskich gier komputerowych* (Warszawa: Open Beta, 2015).

[218] "Rynek komputerowy."

Although Wasiak is right to make the claim that "from the user's point of view there was no major change in the computer market in 1989,"[219] the so-called Wilczek's Act, in force since January 1, 1989 marks the beginning of the period when market processes started to replace state processes in a systematic and official manner. The Act, proposed by Mieczysław Wilczek (then the Minister of Industry) liberalised the law on private business and introduced a laissez-faire system which led to an explosion of small businesses and paved the way for the capitalist transformation of Poland. However, as we know from this chapter, novel computer uses became the foundation for the creative industries, which did not start from scratch after the fall of communism in 1989, but were already being developed in the final years of the PPR.

The diffusion of microcomputers in PPR took place at a time when the communist system was in a state of escalating crisis. At the time when the Iron Curtain was slowly falling apart, the "microcomputer revolution" had reached normal citizens and introduced new cultural practices into their everyday lives. The aim of this chapter has been to present the two main forces that shaped this process. On one hand, we had the authorities and their systemic action, and on the other, extra-systemic initiatives related to the hobbyist movements and emerging fan cultures.

The activity of the state consisted of top-to-bottom and all-in-one initiatives, such as production of microcomputers for the purpose of introducing computer science curriculums to secondary education institutions. In the end, the solutions proposed by the Party proved ineffective, mostly due to the worsening economic conditions in the whole Eastern Bloc, as well as many other shortcomings of the communist bureaucratic machine that resulted in general systemic inertia. The grassroots movements, based on free market principles, turned out to be much more effective in the dissemination of new technologies, and the computer fairs were one of the precursors of the neoliberal capitalism that is still present in today's Poland. However, it is not my goal to evaluate which of the two approaches was superior. It should be noted that all state initiatives were innovative, egalitarian, and in many ways valuable to the development of microcomputing cultures in Poland. The communist authorities understood the importance of computerisation and tried to devise strategies to make it happen. The value of these attempts ought to be recognised, especially in contrast to other new technologies like video, which remained outside of the area of the state's interests until the fall of the system.

Nevertheless, it would be a significant simplification to put these two spheres of activity in complete opposition, one systemic and the other extra-systemic.

---

[219] Wasiak, "'Grali i kopiowali,'" 209.

There was a significant overlap in the activities related to both models of micro-computer dissemination. This was not a phenomenon unique to Poland, since similar processes were playing out in other countries, and not just in the East-ern bloc. We could refer to the history of the informatisation of Finland[220] or Great Britain[221] to see that there were both top-to-bottom and grassroots movements in capitalist conditions, too. Still, the "silicon wave"[222] spreading across the world had to crash against the Iron Curtain on its way to Poland. For this reason, the Polish history of the cultural appropriation of microcomputers is different in its essence to the analogous processes in the countries outside the Eastern bloc. And even within the Comecon countries there were signifi-cant differences in the profile and scope of cultural practices related to the use of microcomputers. As an example, Czechoslovakia saw the production of po-litically committed computer games, which as far as this study shows, was not a Polish phenomenon.[223]

The present research largely confirms the earlier findings of Bartłomiej Klus-ka, Patryk Wasiak, and others. However, it also illustrates that while we know of the most prevalent cultural practices related to the use of microcomputers in the late PPR, there remain many areas still unexplored and many stories wait-ing to be told. Further research could explore unconventional uses of microcom-puters, such as cultural practices not falling under the "playing and copying" par-adigm reviewed by Wasiak. For example, one such area could be digital art not related to the demoscene.[224] In general, not enough attention was paid to the ex-periences of atypical users, as the studies were mostly focusing on young males. As I have already discussed in the methodology, there are complex reasons for this, but hopefully this very book will help in reaching new respondents. Finally, there is a lot of potential for in-depth comparative studies which could involve investigating transnational histories.

---

[220] See Petri Saarikoski, "Computer Courses in Finnish Schools, 1980–1995," in *History of Nordic Computing 3. Third IFIP WG 9.7 Conference, HiNC 2010, Stockholm, Sweden, October 18–20, 2010,* eds. John Impagliazzo, Per Lundin, and Benkt Wangler (Berlin: Springer, 2011).

[221] See Gazzard, "Now the Chips."

[222] See Kluska, Rozwadowski, *Bajty polskie,* 11–20.

[223] See Jaroslav Švelch, "Indiana Jones Fights the Communist Police: Local Appro-priation of the Text Adventure Genre in the 1980s Czechoslovakia," in *Gaming Globally: Production, Play, and Place,* eds. Nina B. Huntemann, Ben Aslinger (Basingstoke: Pal-grave Macmillan, 2013).

[224] See Melanie Swalwell and Maria B. Garda, "Art, Maths, Electronics and Micros: The Late Work of Stan Ostoja-Kotkowski," *Arts* 8, no. 1 (2019): 23.

Acknowledgments

First, I would like to express my gratitude to all the respondents who agreed to be interviewed for the purpose of this research project and who shared their memories about the cultural appropriation of microcomputers in the PPR. Furthermore, special thanks go to all my colleagues and collaborators that patiently awaited this research to see the light of day, many of them contributing their valuable comments to the manuscript, including Bartłomiej Kluska, Jaroslav Švelch, Paweł Grabarczyk and Stanisław Krawczyk. This publication has received additional funding from the Finnish Academy of Sciences and Letters (AKA) as part of the project of the Centre of Excellence in Game Culture Studies (CoE-Game-Cult, [312396]).

Translation: Aleksandra Czyżewska-Felczak and Stanisław Krawczyk

Bibliography

Bałtowski, Maciej, and Szymon Żminda. "Sektor nowych prywatnych przedsiębiorstw w gospodarce polskiej – jego geneza i struktura." *Annales Universitatis Mariae Curie-Skłodowska. Sectio H, Oeconomia* 39, no. 4 (2005).

Bar, François, Matthew S. Weber, and Francis Pisani. "Mobile Technology Appropriation in a Distant Mirror: Baroquization, Creolization, and Cannibalism." *New Media & Society* 18, no. 4 (2016).

Barbrook, Richard, and Andy Cameron. "The Californian Ideology." *Mute* 1, no. 3 (1995). Accessed May 21, 2020. http://www.metamute.org/editorial/articles/californian-ideology

Barelkowski, Matthias. "Hobby bez granic? Rzecz o krótkofalarstwie w Polsce w latach 1925–1990." *Przegląd Historyczny* 109, no. 4 (2018).

Bellows, Anne C. "One Hundred Years of Allotment Gardens in Poland." *Food & Foodways* 12, no. 4 (2004).

Bieszki, Mirosław, and Marian Pianowski. "Luka technologiczna," *Przegląd Techniczny*, no. 44 (1985).

Boukharaeva, Louiza M., and Marcel Marloie. *Family Urban Agriculture in Russia: Lessons and Prospects.* Cham: Springer, 2015.

Cellary, Wojciech, and Paweł Krzysztofiak. "Historia polskiego komputera edukacyjnego," December 22, 2015. Accessed January 30, 2020. www.computerworld.pl/news/404071_1/Historia.polskiego.komputera. Edukacyjnego.html

Ceruzzi, Paul E. *Computing: A Concise History.* Cambridge, MA: MIT Press, 2012.

Coppa, Francesca. "A Brief History of Media Fandom." In *Fan Fiction and Fan Communities in the Age of the Internet*, edited by Karen Hellekson and Kristina Busse. Jefferson, NC: McFarland, 2006.

Drachal, Halina. "Flirt z komputerem." *Głos Nauczycielski*, no. 1 (1987).

Empacher, Adam B. *Maszyny liczą same?* Warszawa: Wiedza Powszechna, 1960.

Fidelis, Małgorzata. *Women, Communism, and Industrialization in Postwar Poland*. Cambridge: Cambridge University Press, 2010.

Florczyk, Andrzej. "Ile komputerów jest w naszych domach." *Komputer*, no. 1, 1989.

Garda, Maria B., and Paweł Grabarczyk. "Technologiczna wzniosłość demosceny." In *Sztuka ma znaczenie*, edited by Dagmara Rode, Maciej Ożóg, Marcin Składanek. Łódź: Wydawnictwo Uniwersytetu Łódzkiego, in print.

Garda, Maria B., and Paweł Grabarczyk. "'The Last Cassette' and the Local Chronology of 8-Bit Video Games in Poland." In *Games and the Local*, edited by Melanie Swalwell, in print.

Gazzard, Alison. *Now the Chips Are Down: The BBC Micro*. Cambridge, MA: MIT Press, 2016.

Graczyk, Roman. *Tropem SB. Jak czytać teczki?* Kraków: Wydawnictwo Znak, 2007.

Hård, Mikael, and Andrew Jamison. *Hubris and Hybrids: A Cultural History of Technology and Science*. New York: Routledge, 2005.

Hughes, Thomas P., Trevor J. Pinch, and Wiebe E. Bijker, eds. *The Social Construction of Technological Systems: New Directions in the Sociology and History of Technology*. Cambridge, MA: MIT Press, 1989.

Informatyka, no. 1, 1986.

IPN Po 06/281/27.

IPN_Sz_00_11_1711.

j.r. "Komputeryzujemy się." *Komputer* no. 8 (1986).

JŁ. *Sargon II* (game review). *TOP* 11 (I), December 18, 1987.

Jachnicki, Tomasz, and Sebastian Stecewicz, "'Komputeryzujmy się.' Wywiad z Ryszardem Kajkowskim," June 13, 2016. Accessed May 19, 2020. https://savethefloppy.com/2016/06/13/komputeryzujmy-sie-wywiad-z-ryszardem-kajkowskim-czesc-i.html

Kaczyński, Jarosław. "Mikrokomputer w lesie." *Przegląd Techniczny*, no. 42 (1985).

Karwicka, Krystyna. "Artyści i rzemieślnicy." Accessed January 30, 2020. http://www.computerworld.pl/news/ 315398/Artysci.i.rzemieslnicy.html

Karwicka, Krystyna. "Hobbyści z konieczności." *Przegląd Techniczny*, no. 41 (1985).

Kirkpatrick, Graeme. "Meritums, Spectrums and Narrative Memories of 'Pre-Virtual' Computing in Cold War Europe." *The Sociological Review* 55, no. 2 (2007).

Kitov, Vladimir, and Nikolay Krotov. "The Main Computer Center of the USSR State Planning Committee (MCC of Gosplan)." *Selected Papers: 2017 Fourth International Conference on Computer Technology in Russia and in the Former Soviet Union (SORUCOM)*, edited by Irena Krayneva and Alexander Tomilin. Piscataway: 2017.

Klawiński, Jerzy. "Jeden pasterz." *Informik*, no. 1 (1987).

Kluska, Bartłomiej. *Automaty liczą. Komputery PRL*. Gdynia: Novaeres, 2013.

Kluska, Bartłomiej. "'Komputeryzacja jakby od końca' obywateli, przedsiębiorstw i uczelni PRL-u." In *High-tech za żelazną kurtyną. Elektronika, komputery i systemy sterowania w PRL*, edited by Mirosław Sikora with Piotr Fuglewicz, Katowice: Instytut Pamięci Narodowej, 2017.

Kluska, Bartłomiej. *PESEL w PRL: informacja czy inwigilacja?*. Łódź: Księży Młyn Dom Wydawniczy, 2019.

Kluska, Bartłomiej, and Mariusz Rozwadowski. *Bajty polskie*, 2nd ed., revised and updated. Self-published, 2014.

Kluska, Bartłomiej, and Mariusz Rozwadowski. *Bajty polskie*. Łódź: Samizdat Orka, 2011.

Kłos, Andrzej. "Rys historyczny rozwoju informatyki w polskiej elektroenergetyce." Paper presented at the conference "50 lat zastosowań informatyki w polskiej energetyce." Warszawa, April 21, 2009. Accessed January 30, 2020. http://apw.ee.pw.edu.pl/sep-ow/PLI/konf/zipe'09/klos/RysHist-InfwEE-AK.htm

"Komputer dla przemysłu." *Przegląd Techniczny*, no. 47 (1986).

Kordiukiewicz, Aleksy. "Cudowne mikrokomputery." *Przegląd Techniczny*, no. 1 (1987).

Koselleck, Reinhardt. *Semantyka historyczna*. Translated by Wojciech Kunicki. Poznań: Wydawnictwo Poznańskie, 2001.

Kosman, Marcin. *Nie tylko Wiedźmin. Historia polskich gier komputerowych*. Warszawa: Open Beta, 2015.

Krawczyk, Stanisław. "Gust i prestiż. O tworzeniu pola prozy fantastycznej w Polsce." PhD diss., University of Warsaw, 2019.

Kulisiewicz, Tomasz. "Polskie komputery 1948–1989. Produkcja i zastosowania na tle geopolitycznym i gospodarczym." In *High-tech za żelazną kurtyną. Elektronika, komputery i systemy sterowania w PRL*, edited by Mirosław Sikora with Piotr Fuglewicz. Katowice: Instytut Pamięci Narodowej, 2017.

Lenoir, Tim, and Henry Lowood. "Theaters of War: The Military-Entertainment Complex." In *Collection – Laboratory – Theater: Scenes of Knowledge in the 17th Century*, edited by Helmar Schramm, Ludger Schwarte, and Jan Lazardzig, Berlin: Walter de Gruyter, 2005.

Liszka, Michał. "Opol-1 i co dalej." *Komputer*, no. 1 (1986).

Loguidice, Bill, and Matt Barton. *Vintage Game Consoles: An Inside Look at Apple, Atari, Commodore, Nintendo, and the Greatest Gaming Platforms of All Time*. New York: Focal Press, 2014.

Łukasik-Makowska, Barbara. "Sprzężenie zwrotne: Mikrokomputery, wystąp!" *Przegląd Techniczny*, no. 20 (1985).

Magier, Dariusz, and Michał Mroczek. "Program 'Awangarda XXI wieku' – w poszukiwaniu nowego modelu pasa transmisyjnego." In *To idzie młodość. Młodzież w ideologii i praktyce komunizmu*, edited by Dariusz Magier. Lublin – Radzyń Podlaski: Archiwum Państwowe, Towarzystwo Nauki i Kultury Libra, 2016.

Majczak, Grzegorz. "Nowości z mikroświata." *TOP*, October 16, 1987.

Majewski, Władysław. "Z czego lepić komputery? Rozmowa z Henrykiem Piłko." *Przegląd Techniczny*, no. 10 (1985).

Mańkiewicz-Cudny, Ewa, and Roman Dawidson. "Kto kocha komputery?" *Przegląd Techniczny*, no. 10 (1984).

Mańkowski, Piotr. *Cyfrowe marzenia. Historia gier komputerowych i wideo*. Warszawa: Wydawnictwo Trio – Collegium Civitas, 2010.

Mańkowski, Piotr. "Polak z Doliny Krzemowej. Wywiad z Lucjanem Wenclem." *Pixel*, no. 1/55 (2020).

Marecki, Piotr, and Tomasz „Tdc" Cieślewicz. *Oni migają tymi kolorami w sposób profesjonalny: narodziny gamedevu z ducha demosceny w Polsce*, Kraków: Korporacja Ha!art, 2020.

Margolis, Jane, and Allan Fisher. *Unlocking the Clubhouse: Women in Computing.* Cambridge, MA: MIT Press, 2002.

Marx, Leo. *The Machine in the Garden: Technology and the Pastoral Ideal in America.* New York: Oxford University Press, 1964.

Mastanduno, Michael. *Economic Containment: CoCom and the Politics of East-West Trade.* Ithaca, NY: Cornell University Press, 1992.

Mazierska, Ewa. "The Politics of Space in Polish Communist Cinema." In *Via Transversa: Lost Cinema of the Former Eastern Bloc,* edited by Eva Näripea and Andreas Trossek. Tallinn: Eesti Kunstiakadeemia, 2008.

Ministerstwo Edukacji Narodowej. *Projekt resortowego programu badawczo-rozwojowego RRI.14 „Informatyzacja procesów dydaktycznych i naukowo-badawczych w szkołach wyższych."* Wrocław, 1988.

Moore, Gordon E. "Cramming More Components onto Integrated Circuits." *Electronics* 38, no. 8 (April 19, 1965).

Motywy, May 20, 1987.

Nye, David E. *American Technological Sublime.* Cambridge, MA: MIT Press, 1994.

Onichimowski, Grzegorz, and Roman Poznański. "Generacja z komputera. Rozmowa z Aleksandrem Kwaśniewskim – ministrem d/s młodzieży." *Bajtek* 4, 1987.

Paśko, Jan Rajmund. "Fotograficzne wydawnictwa seryjne a ruch amatorski w XX wieku." *Annales Universitatis Paedagogicae Cracoviensis. Studia de Cultura* 3, no. 112 (2014).

"Polskie Komputery," accessed January 30, 2020. https://www.facebook.com/polskie-komputery/

Posner, Michael V. "International Trade and Technical Change," *Oxford Economic Papers* 13, no. 3 (1961).

Poznański, Roman. "Drążek sterowy." *Bajtek,* no. 1 (1985).

R.D. [Roman Dawidson]. *Przegląd Techniczny,* no. 40 (1985).

RetroKomp. "Rozmowa z założycielem KOMBI Sławomirem Łosowskim po koncercie na RetroKomp 2016." Accessed January 30, 2020. https://www.youtube.com/watch?v=jsEWsJull5k&t=487s

"Rynek komputerowy." *TOP,* February, 1987.

Saarikoski, Petri. "Computer Courses in Finnish Schools, 1980–1995." In *History of Nordic Computing 3. Third IFIP WG 9.7 Conference, HiNC 2010, Stockholm, Sweden, October 18–20, 2010,* edited by John Impagliazzo, Per Lundin, and Benkt Wangler. Berlin–Heidelberg: Springer, 2011.

Siedlecki, Zbigniew, and Waldemar Siwiński. "RUN czyli zaczynamy." *Bajtek,* no. 1 (1985).

Sikora, Mirosław, and Piotr Fuglewicz, eds. *High-tech za żelazną kurtyną. Elektronika, komputery i systemy sterowania w PRL.* Katowice: Instytut Pamięci Narodowej, 2017.

Singh, Sanjay K. "The Diffusion of Mobile Phones in India." *Telecommunications Policy* 32, no. 9–10 (2008).

Smith-Shomade, Beretta E. "Appropriation." In *Keywords for Media Studies,* edited by Laurie Ouellette and Jonathan Grey. New York: New York University Press, 2017.

Sommer, Vítězslav. "Scientists of the World, Unite! Radovan Richta's Theory of Scientific and Technological Revolution." In *Science Studies during the Cold War and Beyond,* edited by Elena Aronova and Simone Turchetti. Basingstoke: Palgrave Macmillan, 2016.

Sonda, Czynnik Si special, Part 3, min. 8:05. Source: TVP archives.

Stachniak, Zbigniew. "Red Clones: The Soviet Computer Hobby Movement of the 1980s." *IEEE Annals of the History of Computing* 37, no. 1 (2015).

Staszenko-Chojnacka, Dominika. "Narodziny medium. Gry wideo w polskiej prasie hobbystycznej końca XX wieku." PhD diss., University of Łódź, 2020.

Stawowy, Adam. "Komputery w instytucjach." *Przegląd Techniczny*, no. 1 (1987).

Suominen, Jaakko, and Jussi Parikka. "Sublimated Attractions: The Introduction of Early Computers in Finland in the Late 1950s as a Mediated Experience." *Media History* 16, no. 3 (2010).

Švelch, Jaroslav. *Gaming the Iron Curtain: How Teenagers and Amateurs in Communist Czechoslovakia Claimed the Medium of Computer Games.* Cambridge, MA: MIT Press, 2018.

Švelch, Jaroslav. "Indiana Jones Fights the Communist Police: Local Appropriation of the Text Adventure Genre in the 1980s Czechoslovakia." In *Gaming Globally: Production, Play, and Place*, edited by Nina B. Huntemann and Ben Aslinger. Basingstoke: Palgrave Macmillan, 2013.

Švelch, Jaroslav. "Tall Tales and Murky Memories of Computer Gaming in 1980s Czechoslovakia." Filmed October 18, 2019 at the 1st Collaborative Game Histories seminar in Tampere. Video. Accessed January 30, 2020, https://youtu.be/A8OcX-wpMQcY?t=5858, 1:37:38.

Swalwell, Melanie. "Questions about the Usefulness of Microcomputers in 1980s Australia." *Media International Australia*, no. 143 (2012).

Swalwell, Melanie, and Maria B. Garda. "Art, Maths, Electronics and Micros: The Late Work of Stan Ostoja-Kotkowski." *Arts* 8, no. 1 (2019).

Sysło, Maciej M. "Zasługi PRL dla edukacji informatycznej." In *High-tech za żelazną kurtyną. Elektronika, komputery i systemy sterowania w PRL*, edited by Mirosław Sikora with Piotr Fuglewicz. Katowice: Instytut Pamięci Narodowej, 2017.

Świdziński, Jacek. "Krzemowe wyzwanie." *Związkowiec. Tygodnik Popularny*, no. 27 (85), July 6, 1986.

Tok. "Komputer nie śledź," *Świat Młodych*, no. 151 (1986).

Toffler, Alvin. *Trzecia fala* [*The Third Wave*]. Translated by Ewa Woydyłło. Warszawa: Państwowy Instytut Wydawniczy, 1986.

Tokatli, Nebahat. "A comparative report on the profiles of retailing in the emerging markets of Europe: Turkey, Poland, Hungary, Portugal, and Greece." *Journal of Euromarketing* 8, no. 4 (2000).

Turkle, Sherry. *The Second Self: Computers and the Human Spirit*. Cambridge, MA: MIT Press, 2005.

W.M. *Przegląd Techniczny*, no. 42 (1985).

Wacławek, Roland. "Mikrokomputery w natarciu." *Młody Technik*, no. 7 (1983).

Wacławek, Roland. *Z mikrokomputerem na co dzień*. Warszawa: Nasza Księgarnia, 1987.

Wasiak, Patryk. "'Grali i kopiowali' – Gry komputerowe w PRL jako problem badawczy." In *Kultura popularna w Polsce w latach 1944–1989: problemy i perspektywy badawcze,* edited by Katarzyna Stańczak-Wiślicz. Warszawa: Instytut Badań Literackich PAN, 2012.

Wasiak, Patryk. "Playing and copying: Social practices of home computer users in Poland during the 1980s." In *Hacking Europe. From Computer Cultures to Demoscenes*. London: Springer, 2014.

Wasilewska, Karolina. *Cyfrodziewczyny. Pionierki polskiej informatyki*. Warszawa: Wydawnictwo Krytyki Politycznej, 2020.

Wierzbicki, Marek, Jerzy Eisler, Joanna Sadowska, and Łukasz Kamiński. "Młodzież w PRL. Dyskusja." *Pamięć i Sprawiedliwość* 10, no. 1 (2011).

Wierzchowska, Aleksandra. "SF jest ulubioną rozrywką informatyków. Polski fandom a popularyzacja elektroniki." In *High-tech za żelazną kurtyną. Elektronika, komputery i systemy sterowania w PRL*, edited by Mirosław Sikora with Piotr Fuglewicz. Katowice: Instytut Pamięci Narodowej, 2017.

Wróblewska, Agnieszka. "Bilion do podziału." *Przegląd Techniczny*, no. 45 (1986).

Zalot, Grzegorz. "Kluby Mikrokomputerowe: 'Informik.'" *Przegląd Techniczny*, no. 10 (1985).

Interviews

Halina Bednarska, interview by Maria B. Garda, Łódź, January 23, 2017.

Marcin Borkowski, interview by Maria B. Garda, Warszawa, December 4, 2016.

Artur Ciemięga, interview by Maria B. Garda, Wrocław, November 28, 2015.

Tomasz Cieślewicz, interview by Maria B. Garda, Warszawa, November 16, 2015.

Waldemar Czajkowski, interview by Maria B. Garda, Wrocław, May 3, 2016.

Piotr Fuglewicz, interview by Maria B. Garda, Katowice, November 23, 2015.

Paweł Grabarczyk, interview by Maria B. Garda, Łódź, February 16, 2014.

Tomasz Grochowski, interview by Maria B. Garda, Warszawa, November 15, 2015.

Grzegorz Juraszek, interview by Maria B. Garda, Racibórz, September 21, 2014.

Wojciech Mikołajczyk, interview by Michał Sieńko, Lublin, November 28, 2015.

Wojciech Nowak, interview by Maria B. Garda, Racibórz, September 21, 2014.

Barbara Ostrowska, interview by Michał Sieńko, Lublin, November 28, 2015.

Zbigniew Rudnicki, interview by Maria B. Garda, Katowice, November 23, 2015.

Paweł Sikorski, interview by Maria B. Garda, Warszawa, November 28, 2016.

Arkadiusz Staworzyński, interview by Maria B. Garda, Łódź, November 22, 2014.

Roland Wacławek, interview by Maria B. Garda, Katowice, December 4, 2016.

Part IV

Krzysztof Jajko

TELEVISION FROM THE SKY
FOR EVERYONE

Cosmic television in Poland[1]

Adhering strictly to factual material, the year 1975 could be considered the beginning of satellite television history in Poland. One of the effects of Poland joining the international organisation of socialist countries (Intersputnik) was that a satellite ground station had begun operating in Psary, an area in the Świętokrzyskie Mountains. It must be stressed that in that era, regardless of which side of the Iron Curtain you were on, satellite communication relied on the same scheme of operation. Telecommunication satellites were the basis of communications, and the TV programmes they transmitted were sent out somewhat incidentally. What is most important, however, is that the transponders placed on the telecommunication satellites had very low power (20 W), which necessitated the use of parabolic antennas of great diameter in order to receive audiovisual signals (the first dish installed in Psary measured 12.5 metres, the second one 32 metres).[2]

These purely technical limitations meant that for a long time, TV satellite communications were limited to communication between large TV networks. In the case of Poland, it was Radiokomitet – the monopolist in TV programme broadcasting – that had the best access to space signals. Because that institution was under the direct control of the Party, even more so than most, it is no exaggeration to say that the state had a monopoly over satellite television in its first years of functioning in the PPR. It was the government that decided which foreign programmes were suitable for distribution in socialist Poland, and they were retransmitted through Polish Television's broadcast network as part of the first or second channel's listings. It is important to note that the Party was extremely cynical in this regard, as best illustrated by the minister of communications, Edward Kowalczyk's remarks during his visit to Psary station in 1975: "[...] the modern man can be free thanks to the access and use of information."[3]

---

[1] Subchapters *Cosmic television in Poland*, *The importers* and *The Antenna Guys* have been published previously (in Polish) in "Skrzynki, talerze i transformacja. Młodzieńcze lata telewizji satelitarnej nad Wisłą," *Panoptikum*, no. 15 (2016): 30–52.

[2] Józef Dolecki, "Centrum łączności satelitarnej w Psarach," *Audio-Video*, no. 1 (1985): 10–11.

[3] Kazimierz Michalewicz, "Ziemia, kosmos, ziemia. W stacji satelitarnej w Psarach," *Studio*, no. 5 (1975): 32.

It was a very specific kind of freedom, however: a citizen could access and use only the information which had been previously vetted by a suitable body of Party experts.

Fig. 1. The ground satellite station in Psary

Source: KAW-CAF

This practice of regulating satellite TV access persisted until the end of the PPR. The best proof of this is *Bliżej świata (Closer to the World)*, a show extremely popular in the late 80s, hosted by Jerzy Klechta. In it, Polish viewers were served excerpts from Western television, music videos, and other such things. News editorial boards were eagerly using foreign audiovisual materials as well. With this in mind, it can be argued that in the 70s and 80s watching satellite TV for many Poles was equivalent to watching Western audiovisual materials retransmitted by Polish Television as parts of domestic programmes. Satellite TV in a socialist country was, therefore, a complete distortion of the ideas underlying the spread of satellite television in the West. In the West, the new medium of cable television was used to significantly diversify the programmes on offer, but in Poland it was used only to slightly expand domestic TV listings, with strict elimination of any politically or morally incorrect materials.

This does not mean, however, that the communist authorities had a monopoly on satellite television until the very last days of the PPR. By the early 80s, satellite sets had already appeared in the West, making it possible to directly receive space signals broadcast by telecommunication satellites, and to watch them without the cable networks' mediation. In this way, individual reception of satellite TV was made possible in Europe long before the TV broadcasters had officially decided to broadcast TV programmes directly to private viewers (DBS – direct broadcast satellites). The news about free satellite TV quickly found its way to Poland, as had been happening with other media novelties. There is evidence to suggest that the first satellite TV sets were being secretly imported by the end of 1983. Equipped with antennas with a 180 cm diameter, these sets easily allowed for the reception of signals from Western telecommunication satellites like Eutelsat or Intelsat. As a result, an appetite for the unconstrained choice offered by satellite TV broadcasters began to develop in Poland. From that moment on, the reception of space signals was happening in two ways. While a great majority of Poles was condemned to selected programmes like *Closer to the World*, a smaller lucky group, was feasting their senses on a continuous stream of pictures and sounds from the West.

The practice of private importation of receivers and satellite TV antennas to Poland provoked a time of satellite anarchy, as had been the case with VCRs as well. However, the Party quickly realised that it could not allow for a situation where it would lose control over yet another sector of audiovisual communication. While the VCR was a domain of escapist entertainment,[4] the satellite TV threatened with an influx of revolutionary, purely political

---

[4] An exception from this rule were videotapes published in the second circulation, propagating forbidden materials on political topics. Such non-cinema distribution was organised for the renowned movie *Przesłuchanie* (*Interrogation*), among others. Interviews with members of the Silesian Science Fiction Club, conducted as part of the New Media project, indicate that illegal screenings of this movie were taking place not only in the opposition circles, but also among science-fiction fans who were not engaged in politics. This indicates that tapes with revolutionary content could have been reaching considerable sections of society. However, considering the small number of titles that

content in Western news programmes, and posed a greater threat. The final straw was the news that by the end of the 80s, Western countries wanted to have installed DBS satellites in European skies, which, thanks to the 45 W transponders, made it possible to receive TV programmes even with 60 cm diameter antennas.

Only one question remained for the Party: how to deal with the unwanted child of space communication? A book published by the Ministry of Defense in 1987 titled *Gość czy intruz z kosmosu (A Guest or an Intruder from Space?)* offers some insight into the communist authorities' dilemma over this matter. Although the authors discuss the two options listed in the title for almost 200 pages, the final conclusion of this dilemma is clear: if domestic television could not offer an attractive alternative to the Western programmes, which exploded with vivid colours and sounds, a non-cultural counterattack should be mounted. Such a confrontational stance is fully justified, the publication's authors argued, writing "we have no doubt about the Western propaganda's evident interest in the DBS system, with the intention of using it in an ideological fight against countries of different political systems."[5]

It is difficult to ascertain when, exactly, those words were written. However, taking the publishing cycle into account, it can be assumed that these ideas came into being at the beginning of 1986. On February 20 of that year, the Minister of Communication published the decree regulating the right to own "receivers other than of common use." The ministry's unprecedented decision was explained in *Polityka* by Janusz Fajkowski, director of the Department of Radiocommunication Service in the Ministry of Communication:

> No country is indifferent to the issue of foreign satellite system reception. This cannot be allowed to take over. What would our rooftops look like, if everyone installed an individual antenna? Some spatial order has to be kept in the landscape. Besides, there are also formal and legal issues, and most importantly the issue of payment for using foreign satellite systems.[6]

Like most communist authorities' statements, this one too was characterised by a high degree of hypocrisy. Indeed, the necessity of introducing permits to own and use satellite TV was justified by pointing out problems that were not really anyone's concern.

---

were published in the second circulation, the scale of the phenomenon was surely much smaller than in the case of "pirate" distribution of entertainment movies from the West.

[5] Andrzej Fryszkiewicz, Marian Grabski, Janusz Sarosiek, *Gość czy intruz z kosmosu?* (Warszawa: Wydawnictwo Ministerstwa Obrony Narodowej, 1987), 71.

[6] Marek Henzler, "U wrót satelitarnego raju," *Polityka*, May 3, 1986: 10.

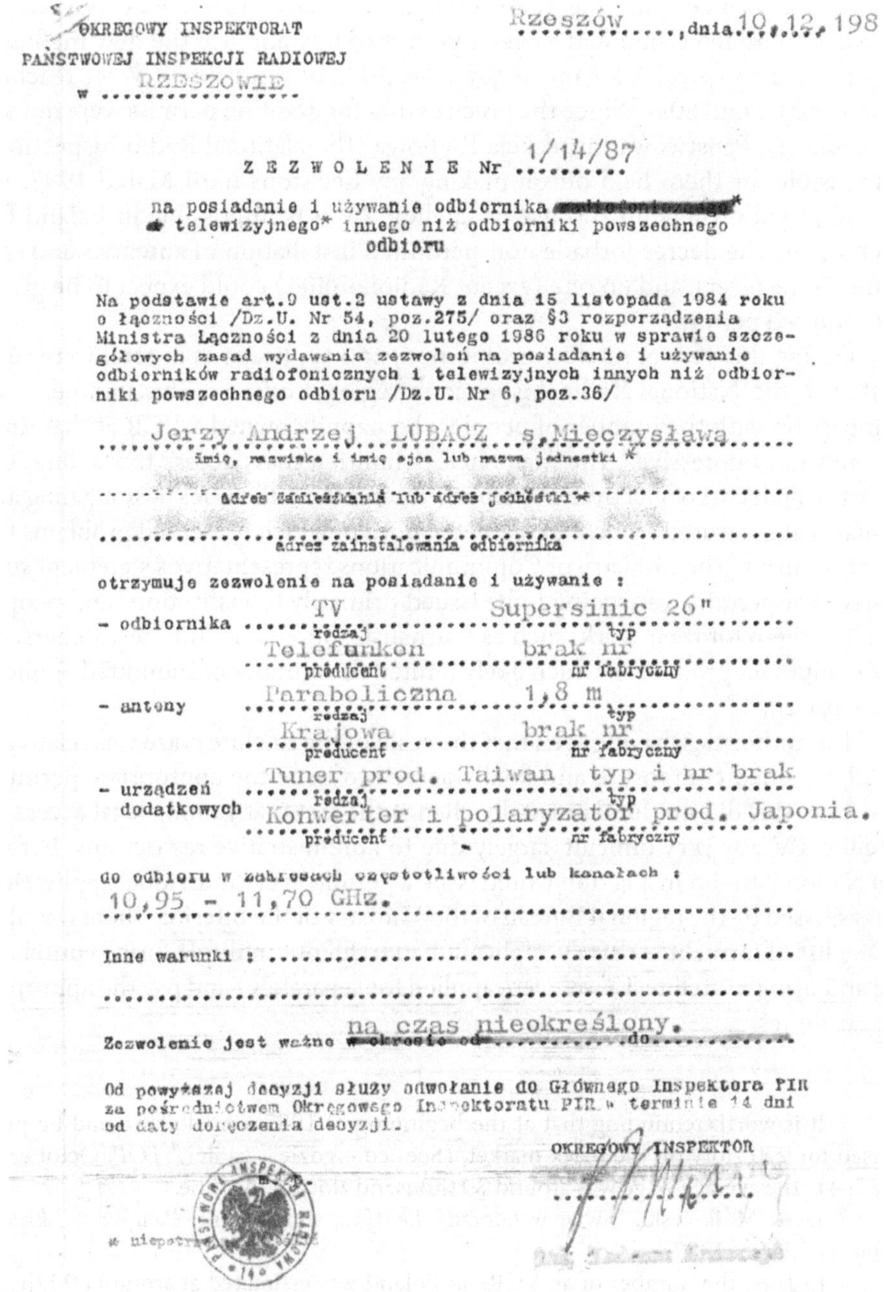

OKRĘGOWY INSPEKTORAT

PAŃSTWOWEJ INSPEKCJI RADIOWEJ

w RZESZOWIE

Rzeszów, dnia 10.12. 198

Z E Z W O L E N I E Nr 1/14/87

na posiadanie i używanie odbiornika ~~radiofonicznego~~*
* telewizyjnego* innego niż odbiorniki powszechnego
odbioru

Na podstawie art.9 ust.2 ustawy z dnia 15 listopada 1984 roku
o łączności /Dz.U. Nr 54, poz.275/ oraz §3 rozporządzenia
Ministra Łączności z dnia 20 lutego 1986 roku w sprawie szcze-
gółowych zasad wydawania zezwoleń na posiadanie i używanie
odbiorników radiofonicznych i telewizyjnych innych niż odbiorniki powszechnego odbioru /Dz.U. Nr 6, poz.36/

Jerzy Andrzej LUBACZ s.Mieczysława

imię, nazwisko i imię ojca lub nazwa jednostki *

adres zamieszkania lub adres jednostki*

adres zainstalowania odbiornika

otrzymuje zezwolenie na posiadanie i używanie :

- odbiornika ... TV ... Supersinic 26"
.... rodzaj .... typ
Telefunken ... brak nr
.... producent .... nr fabryczny

- anteny ... Paraboliczna ... 1,8 m
.... rodzaj .... typ
Krajowa ... brak nr
.... producent .... nr fabryczny

- urządzeń ... Tuner prod. Taiwan typ i nr brak
dodatkowych .... rodzaj .... typ
Konwerter i polaryzator prod. Japonia.
.... producent .... nr fabryczny

do odbioru w zakresach częstotliwości lub kanałach :

10,95 - 11,70 GHz.

Inne warunki : ....

....

na czas nieokreślony.

Zezwolenie jest ważne ~~w okresie od .... do ....~~

Od powyższej decyzji służy odwołanie do Głównego Inspektora PIR
za pośrednictwem Okręgowego Inspektoratu PIR w terminie 14 dni
od daty doręczenia decyzji.

OKRĘGOWY INSPEKTOR

* niepotr

Fig. 2. A National Radio Inspection permit issued for ownership
of a satellite television set

Source: Jerzy Lubacz's private archive

In accordance with the ministerial decree, any citizen who wished to own a satellite television had to get a permit to buy and use the new medium, and needed to spend a lot of money (the price of a satellite TV set reached $1,500 in the mid 80s).[7] Since the precise rules for granting permits were not set immediately, Państwowa Inspeckcja Radiowa (the National Radio Inspection) responsible for them held off on making any decisions until March 1987. As a result, legal ownership and use of satellite TV was impossible in Poland for over a year. The decree forbade non-permitted installation of antennas and satellite TV receivers, and no one (except Radiokomitet) could expect to be given the required permit.

During this year of satellite prohibition, 260 applications in total were submitted to the National Radio Inspection's regional offices.[8] This number – in comparison with the number of people who happily owned a VCR at that time – is not too impressive.[9] The main factors limiting mass access to satellite TV were the high cost of electronic components and the difficulties in smuggling antennas that measured 180 cm in diameter, both of which presented problems for potential users. The Ministry of Communications representative's statement suggested that permits were going to be issued primarily to institutions and people needing them for their work, such as journalists, popular culture researchers, or TV equipment producers, which likely limited the number of submitted applications as well.[10]

The authorities, having seen that the scale of the satellite craze was relatively small, regained composure and finally agreed to issue the appropriate permits. However, this did not fundamentally change the fact that gaining legal access to satellite TV was very difficult, largely due to administrative restrictions. Before the National Radio Inspection would issue a permit, every individual application was assessed by the regional Bureau of the Ministry of the Interior. There was also a long list of formalities: every applicant had to fill out multiple forms (purchasing and using a satellite TV set were applied for separately) and pay the appropriate tax duties.[11]

---

[7] It is worth reminding that at the beginning of 1987, one dollar could be purchased for 970 złoty on the black market. (See "co – gdzie – za ile?," *TOP*, October 9, 1987: 4). The average wage was around 30 thousand złoty at the time.

[8] Jacek Mojkowski, "Świat w talerzu," *Polityka*, supplement: *Polityka – Eksport – Import*, March, 1987: 16.

[9] In 1987 the number of all VCRs in Poland was estimated at around 600 thousand. See Krzysztof T. Toeplitz, "My i nasz system," *Polityka*, June 20, 1987: 1, 7.

[10] Henzler, "U wrót."

[11] Jerzy Machejek, "Odmówić, odmówić, odmówić...?," *TV-Sat-Magazyn*, March 15, 1989: 7.

These procedures, which undoubtedly slowed down the expansion of satellite TV in Poland, quickly became the subject of attacks from more progressive commentators, especially those centered around *Polityka*. The biggest advocate for free access to space communication was Krzysztof Teodor Toeplitz, who also had no difficulty in supporting the politics of Wojciech Jaruzelski, the First Secretary of the Polish United Workers' Party from 1981 to 1989. This quite equilibristic stance from one of the main writers of the PPR raises suspicions that this entire journalistic battle for satellite abolition was rigged from the start. At any rate, Toeplitz – appalled mostly by the absurdities of the bureaucratic machine – pointed out, across multiple articles, the necessity of rejecting what he called the "separatist approach."[12] In the commentator's opinion, the time of any government's monopoly on information was coming to an end as a result of the media revolution taking place. To give his considerations adequate weight, Toeplitz appealed to an authority from the highest echelons of power:

> This process, caused by technology, collides in a most fortunate way with the period of mental change brought on to our alignment by Gorbachev's term "openness" reality, in favour of a true reality, and the ambition to shift the information system from a restrictive model relying on suppressing unfavourable or unsuccessful information to a polemical model, one that confronts beliefs while respecting the essential matter of information, treated as open.[13]

Toeplitz demonstrated a substantial gift for exegesis in this passage. It is in no way an overinterpretation, either; on the very first page of his manifesto *Perestroika: New Thinking*, Mikhail Gorbachev wrote in reference to international relations: "We must meet and discuss. We must tackle problems in a spirit of cooperation rather than animosity."[14] With less finesse but much more ardour, Jacek Mojkowski also requested free access to satellite TV in *Polityka*. As he writes in his article from March 1987: "To have to explain in this day and age, why one wants to watch TV is, after all, some nonsense."[15]

In the last months of the PPR, the communist authorities aligned with the views of the press. On February 13 1989, the Minister of Transportation, Maritime Affairs and Communications issued a decree that practically abolished all existing limitations for accessing satellite television. From that point on, anyone who wanted to receive TV programmes from space only had to register their

---

[12] Toeplitz, "My i nasz."

[13] Ibid.

[14] Mikhail S. Gorbachev, *Perestroika: new thinking for our country and the world* (New York: Harper & Row, 1987), 9.

[15] Mojkowski, "Świat."

receiver with the head of the appropriate post office. This, however, did not happen without complications. Although the decree went into effect in early March, the National Radio Inspection's General Inspectorate sent a telex to all its subsidiaries instructing them to suspend responsibilities pertaining to registration between the 1 and 30 of April.[16] As a result, it is difficult today to point to the exact date when the era of free satellite TV access in Poland began.

It seems that a change in the communist authorities' approach toward satellite television happened alongside the development of a satellite counteroffensive plan in the Comecon. In line with the arrangements already made in the year 1987, the first socialist broadcasting satellites were to appear in geostationary orbit between 1990 and 1992; construction of a number of ground broadcasting stations in Poland, Romania, Czechoslovakia, and other countries was planned between 1990 and 1993; and satellite TV receivers for individual use were to be made available to the general public by 1992.[17] These far-reaching plans were ended by the sudden disintegration of the Soviet bloc in the second half of 1989. Of course, knowing the weaknesses of the planned socialist economy, it is likely that the satellite counteroffensive of the Comecon countries would have taken much longer than the officially announced work schedule assumed.

The importers

The problem of individual satellite TV reception in the PPR surfaced for the first time in a public forum when the Minister of Communication's 1986 decree saw the light of day. The first satellite TV reception sets had already appeared in Poland around the end of 1983, and so for over two years owning a satellite television was part of the private domain in socialist Poland. However, the press's total lack of interest in the first wave of satellite fervour, makes describing the early years of the new medium's expansion in the PPR difficult, as we can only really rely on interviews with people who witnessed these events. Below, I will try to describe the first years of satellite television in Poland based on Marek Czajkowski's recollections, as he began importing the first satellite TV sets to Poland just after the end of martial law.[18] Because the practice of so-called private import flourished on an unprecedented scale in the 80s, it is impossible to determine who was the first to smuggle a satellite TV set into Poland, and how many such devices were brought to the country before 1986.

---

[16] MAJ, "Wolno, wolno, wolno," *TV-Sat-Magazyn*, April, 1989: 6.
[17] A. Jamiłowski, "Program z satelity," *Pan*, October, 1987: 46, 60.
[18] Marek Czajkowski, interview, September 16, 2015.

In Marek Czajkowski's case, the idea of importing satellite receivers from behind the Iron Curtain went beyond the small scope of private import. During
the period of martial law he and a Polish friend residing in Sweden had already
decided to open a private company, Porion, in Scandinavia. Originally the company was supposed to import various appliances, equipment, and even furniture,
from Poland to Sweden. The initial plans were quickly changed, however, and in
the years that followed the company specialised exclusively in importing antennas and satellite receivers, but in the opposite direction:

> It was an accident. This business partner of mine went to an electronic trade fair
> in Stockholm and at that fair he saw [satellite TV] for the first time. It was a novelty in
> Sweden as well, after appearing earlier in America. At the end of the exhibition
> he said to me, "There is a company exhibiting satellite antennas. Maybe we could
> get involved." I liked that a lot. I came to Stockholm the next day and bought all
> three satellite antennas that were on display at the fair. I remember paying around
> $10,000 in total. This was a lot of money back then. You could buy six Fiats 125p
> for $10,000. Anyway, I brought those antennas back to Poland.[19]

For Marek Czajkowski, importing satellite TV equipment to Poland did not
pose much of a problem: in the first half of the 80s no one, including the customs
officers, knew what those devices were really for. The success of the entire operation was aided by the fact that the antennas bought at the Stockholm trade fair
had a specific build. The dishes were not made of solid material, but of a netting.
This made the equipment easy to transport across the border, but quickly turned
out to be a major problem:

> After we brought those antennas to Poland, we were trying to catch some signal
> on them for three months, but it was impossible. First of all, Poland's satellite
> range was inadequate and secondly, these antennas had a different converter fre-
> quency.[20]

Undeterred, Czajkowski immediately began familiarising himself with
the new medium and determining what technical parameters the devices must
have in order to operate in Poland. He sums up this period of research as follows:
"It was a great adventure. It took around half a year to work it all out."[21]

Once it was determined what kind of receivers (so-called boxes) and antennas were suitable for satellite TV reception in Poland, Czajkowski imported
the first set for his own use. Satellite TV, as opposed to VHS, was a medium whose

---

[19] Ibid.
[20] Ibid.
[21] Ibid.

ownership was difficult to hide because of the necessity of installing a large, par-abolic antenna (in 1980s Poland, good signal quality was ensured by 180 cm di-ameter dishes or bigger).

> I was living on the sixth floor at the time, and I installed the antenna on the roof-top. I didn't ask anyone for permission, I just put it there. A police officer living in the next building began sniffing around. He wanted to know what it was, where it came from and why it was like that. Later I heard rumours that I'm supposed to be some senior officer and the antenna is for receiving special data. They didn't even know it's for television. And [they said]: "Let's leave it alone."[22]

Establishing whether the authorities from the Ministry of Interior really ig-nored Czajkowski's unauthorised installation of a satellite antenna goes beyond the scope of this chapter. There is no doubt, however, that during the initial years the problem of satellite TV practically did not exist for the communist authorities. It is most clearly illustrated by the fact that the Minister of Communication's de-cree dates only from the beginning of 1986. Likewise, it is clear from Czajkows-ki's recollections that for a long time he was able to conduct his activity without major obstacles. Moreover, the unregulated status of satellite TV was working to the enterprise's advantage.

> When transporting this equipment across the border, you could just say that it's a satellite TV receiver. It wasn't on the list of taxable goods, and so, the customs duty was zero. At the time, no one imagined that there could be something like this, and so luckily, it didn't fall under any regulations. There was no problem in trans-porting it at all, no reason to try to cheat or hide anything.[23]

It is worth noting that Czajkowski tries to present his own interpretation of the authorities' weird, defensive stance:

> It was funny, because this was illegal, after all. Those were communist times. But because the communist authorities had problems already, this subject was brought up before the Central Committee, I found out. Still, the comrades waved it away: "No, this is not worth thinking about. People don't have anything to eat, food prices are rising, and we're discussing antennas for billionaires."[24]

On the one hand, there is a lot of merit to this explanation; on the other hand, it seems that the Party turning a blind eye to the illegal import of satellite TV sets

---

[22] Ibid.
[23] Ibid.
[24] Ibid.

came from the socialist system's general weakness in adapting to ever-changing conditions (both internal and external). After all, such a phenomenon as personal satellite TV could not have been included in the planned economy of the PPR in 1983–1985, because the basic guidelines for economic development in the following years were created when the new medium did not exist yet. We should remember too, that the plan officially accepted by the Comecon for socialist satellite television development in Eastern European countries was meant for the years 1990–1993.

There is no doubt that the "planned delay" written into the socialist economy was incredibly beneficial for companies like Porion. Even though the 80s were a time of thorough economic reforms aimed at making the system "more real" and liberating it in an economic sense, the dogma of restraining competition that a more robust private sector could pose to the state companies was still irrefutable. A situation where a cooperative of craftsmen or artisans would make more of a certain product than a big state facility specialising in it was inconceivable until 1989. A similar situation was being faced in trade and all kinds of services. Therefore, the only areas where private businessmen could prove themselves were economic niches, resulting from the stagnancy of the country's development schedule.[25]

After Marek Czajkowski installed a satellite antenna in his building and made the programmes available to his neighbours through the AZART network (Collective Integrated Radio-Televsion Reception System), his company started gaining recognition. Customers hungry for space signal came from two distinct social circles:

> Most of the customers were private businessmen. People were running small companies and they could afford buying an antenna for $2,000. We also had artists, actors and musicians visiting us, because there were music programmes on the satellite.[26]

What is most striking about this statement is the fact that the satellite television market in Poland between 1983 and 1987 was completely closed; it consisted of private businessmen importing equipment unavailable in the country for other private businessmen. Of course, this situation reflected broader social processes. Since the peak of shortages for consumer goods offered in the official circulation occurred in the 80s, those who had any financial surplus thanks to private economic activity were forced to find other ways of investing the generated capital. As Jerzy Kochanowski states, the non-socialised activity "gave substantial, often untaxed and undisclosed income, invested in black market currency

---

[25] See Dariusz T. Grala, *Reformy gospodarcze w PRL (1982–1989). Próba uratowania socjalizmu* (Warszawa: Wydawnictwo TRIO, 2005), 260, 268–269.

[26] Marek Czajkowski, interview, September 16, 2015.

and gold, or encouraged the development of smuggling techniques, providing both the aforementioned treasured goods as well as luxury items unavailable on the official market."[27]

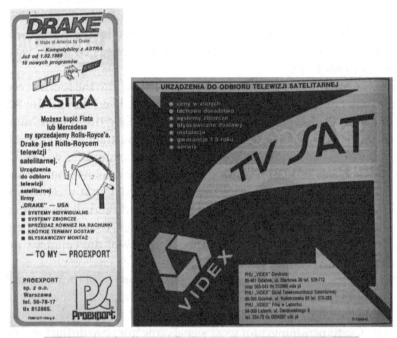

Fig. 3. Advertisements of selected TV SAT equipment providers printed in *TOP*

Source: *TOP*, June 24, 1988: 24; TOP, February 10, 1989: 13; TOP, October 14, 1988: 22

---

[27] Jerzy Kochanowski, *Tylnymi drzwiami. "Czarny rynek" w Polsce 1944–1989* (Warszawa: Wydawnictwo W.A.B, 2015), 20.

This narrowing of the target group of satellite TV clients to private entrepreneurs gave rise to certain advantages from a marketing point of view. Since the businessmen active in the non-socialised sector were a rather uniform community bound by close personal relations, news of a new medium spread without the need for any advertisement. This was important because before the appearance of the first magazine specialising in publishing advertisements, 1987 launched *TOP*, it was virtually impossible to announce the sale of satellite TV receiver sets in the public space:

> You found out on the grapevine. One person told another. You had to have a lot of cash to afford a satellite antenna, so only people affluent for that time were buying them. And if someone who had installed such a television invited a few guests round, well, that created five new customers. Then those five brought another five. That's how it went.[28]

The effectiveness of this bottom-up advertisement campaign can be seen in the fact that the Porion company very quickly started selling around 10-15 satellite TV sets per month. In this way, around 100 receivers and antennas found their way into private hands in the space of a year. If we remember that 260 applications for a satellite TV reception permit were made to the National Radio Inspection between 1986 and 1987, it can be assumed that a lot of them were submitted by Porion's clients.

Of course, it goes without saying that trading satellite TV equipment took place in specific conditions. Since these electronic components were beyond the grasp of socialist technical development, the only option available was importation from Western countries. As a result, the złoty lost all its value in satellite transactions. To own a satellite TV receiver, one had to have exchangeable currency, and a lot of it. This was especially true if one wanted to buy really good equipment, such as the Japanese MASPRO:

> It had only one drawback: the price. Obviously, everything that's Japanese is certainly more expensive. And a typical MASPRO set could cost even close to $2,000. This was big money back then. Nowadays, you get something like that for free.[29]

The high cost of satellite TV deterred few who could afford it, as the new medium brought about an incredible state of social excitement:

---

[28] Marek Czajkowski, interview, September 16, 2015.
[29] Ibid.

There was so much enthusiasm back then. Enthusiasm about suddenly having access to information from the wider world. After all, there was only *Trybuna Ludu* and *Życie Warszawy* and the first and second channel of state television.[30]

Because of this enthusiasm, not only businessmen were drawn to Porion. Enthusiasts of new media technologies from around Poland quickly arrived at its door. Their goal was not only to come into the possession of a satellite TV receiver set, but also to promote satellite television outside of Warsaw. Czajkowski quickly realised that enthusiasm for the medium could be utilised by turning customers into satellite TV installers. Therefore, he quickly started courses allowing future fitters to learn the basic technical aspects of satellite television:

> We didn't deal with installation as a company. We did, however, train people interested in satellite TV from around Poland. They didn't have access to antennas, equipment, or even a concentric cable, so they came to us and we taught them about everything. The benefit was that when they were installing the antennas, the antennas came from us, so as a result we were selling a lot more.[31]

Those first "professional" installers came from different cities, including Cracow, Gdańsk, Turek and Łódź. Among them was Jerzy Lubacz from Mielec. From his success in southern Poland, one can glean that Czajkowski was right to think the satellite enthusiasts would turn a profit. Lubacz's comments below, describe the scale of his activity in the late 80s, and to what degree his work contributed to the spread of knowledge about satellite TV among Poles:

> I wanted to keep working with this company [Porion], to gain the appropriate skills, get the equipment from them and install it for people. In this kind of job, the beginning is most difficult though, when no one knows anything. So, I had the idea to connect the buildings here. A friend from the service who was dealing with AZART helped me […]. We made inserts for a separate special channel, plugging in just one at first. Then I bought the second piece of equipment and plugged in a second channel, connecting six buildings one by one.
> […]
> After buying this antenna and doing this experiment, I organised the so-called displays of satellite equipment and its reception, to promote what I'd been doing. I exhibited this in the cultural centre in Tarnów's Azoty [industrial facility], in Rzeszów, and in Boguchwała […]. After these promotions, the orders came in immediately. The first order I had was for an installation in Poznań – a foreign company. Someone

---

[30] Ibid.
[31] Ibid.

reached out to me through a newspaper. [...] Later on I did them in Cracow, in Brz-esk, and in Tarnów. There were lots and lots afterwards. [...] Plenty of people would come to my home [as well], to see how it all worked.[32]

Over the years, competition got fiercer, as other satellite TV set suppliers appeared alongside Porion in Poland. In 1988, *TOP* magazine listed the follow-ing companies as providing satellite television: SATELIT, PULSON, VIKOR, and SERVISCO.[33] By looking at the advertisements in the same newspaper we can observe that in the following years, further businesses joined this still narrow elite. Examples include ARCON, operating in Warsaw, DIGITAL from Gdańsk, and ROMATEC, operating in Zielona Góra. The situation for these "second wave" importers was infinitely better than that of the pioneers like Porion. They could count on an additional channel of communication with customers, name-ly the press. Indeed, apart from the aforementioned *TOP* magazine, the monthly *TV-Sat-Magazyn* launched at the beginning of 1989. As its title attests, it was dedicated to satellite television exclusively and targeted both existing customers and people planning on buying satellite equipment.

From the beginning, the new companies relied on the organisational expe-rience of Porion, utilising the business model of creating branches which dealt with installing and selling equipment. As evidenced by the advertisements print-ed in the *TV-Sat-Magazyn* in 1991, the DIGITAL company had 14 branches around the country, and the firm ROMATEC had as many as 43. One of them was a company owned by Jerzy Lubacz.

The vigorous growth of competition forced Marek Czajkowski to change his business strategy. Indeed, the next chapter in Porion's history was not individ-ual satellite television, but cable TV, which it first started installing in Warsaw's neighborhood of Żoliborz. By the end of the 90s, Czajkowski's company had spread "cable" across almost the entirety of Warsaw. It also created a local televi-sion station for its subscribers, called Porion TV. But that is a topic for another article. Here, I would only like to make a few comments on some of Marek Czaj-kowski's statements.

First, attention must be paid to the phenomenon of enthusiasm, which sat-ellite television elicited from Poles. The group of proud satellite TV set owners may have been small in absolute numbers, but the phenomenon of watching un-censored satellite television took place on a larger scale. Certainly, as Czajkows-ki mentions, group reception was common practice. He himself had frequently received acquaintances interested in space transmission, and even – like Lubacz – shared it through the AZART network with his neighbours. In this sense, even

---

[32] Jerzy Lubacz, interview, March 3, 2015.
[33] "Inwazja z kosmosu," *TOP*, December 25, 1987: 6.

if the government had succeeded in discouraging satellite TV ownership through permits, it could not fully regulate people's access to it. Besides, it seems that the main factors limiting the spread of satellite television in Poland were not political or legislative matters, but economic ones. With the emergence of numerous companies specialising in selling satellite TV sets around 1989, the prices of those devices started dropping due to increased competition. To attract more customers, the importers started selling receivers and antennas for the Polish złoty. As a result, satellite TV became accessible not only to private business owners, but "regular" citizens as well.

Marek Czajkowski's observation, where he connects satellite enthusiasm with a sense of freedom, is also important. However over the top it sounds, satellite television – through in no way a mediated, direct contact with the West – really gave a sense of liberation from the totalitarian country's oppression. That was, incidentally, its advantage over VCR, which I would position lower in the hierarchy of "emancipatory" media. This is confirmed by the following statement from Czajkowski:

> Above all, I got Filmnet. It was an interesting channel, because it played soft porn, on Wednesdays, I think. That was a huge attraction. The VCRs were there already, and people were watching it, but I had it from space.[34]

At first glance, soft porn and the grand term "freedom" do not seem to align. But if we look at the matter from a broader perspective, in the way John Fiske or Michel de Certeau might, such cultural practices, immersed as they are in trivial reality, can hold great potential.[35] Furthermore, as Marshall McLuhan proved a long time ago, the content we receive and the tool through which we receive it are equally important. The communist authorities tried to spread the belief among Poles that by watching programmes such as *Closer to the World* they were experiencing satellite television. In reality, they were exclusively receiving programmes recorded and retransmitted from satellite TV. There was no media freedom to speak of in this case; freedom was the sole privilege of those who had gained access to direct signal from space.

The last matter is the "coincidence" mentioned by Czajkowski, which kick-started all of Porion's future activities. Of course, the fact that his partner secured the satellite reception equipment at the Stockholm trade fair can be seen as a stroke of luck. However, overwriting this coincidence was a broader scheme

---

[34] Marek Czajkowski, interview, September 16, 2015.

[35] See John Fiske, *Understanding Popular Culture. Second Edition* (New York: Routledge, 2010); Michel de Certeau, *The Practice of Everyday Life*, trans. Steven Rendall (Berkeley: University of California Press, 1984).

of circumstances which must be considered favourable to the organised importation of satellite TV receivers into Poland. The trigger was people's foreign connections. Only people who had contacts abroad could have taken the role of those introducing the new media technology in Poland. Probably the best example is the next subject of this chapter, namely Zdzisław Żniniewicz, also known as "Zdzisław the Precise" or "The Antenna Guy."

The Antenna Guys

It sounds unbelievable, but it is true: at the end of the 80s, Poland grew into a European antenna-production potentate overnight. This strange turn of events was thanks to an eccentric craftsman from Szczecin, who had previously produced – among other things – plastic lighters, and the famous "conference spectacles" (which were even used by the government press officer, Jerzy Urban).

His career was turned around by satellite dishes. As a reporter for *Głos Szczeciński* wrote in 1987, the dishes:

[B]rought him fame, not to mention money. After being exhibited at last year's fair in Poznań, they just exploded.[36] The scent of the big world had wafted over. But progress is very quick. If one is just one month late, he is out of the market. Żniniewicz and all those who co-operated with him were not late, however.[37]

One of these collaborators was Adolf Bogacki, who in subsequent years would produce as many satellite dishes on his own as Żniniewicz. Unfortunately, Zdzisław Żniniewicz, the *spiritus movens* of the satellite dish business, died in 2007, but we can better understand Szczecin's satellite miracle by examining an interview with him, which will supplement Adolf Bogacki's story:

A customer says to me one day: "Listen, Dzisiu (he has a funny way of pronouncing my first name and doesn't even attempt to pronounce my surname), I've got a job for you. You're going to make antennas. Satellite dishes. It's a world trend you know and if you start soon, we'll be in the lead." He kept persuading me, encouraging; he came here twice in fact, and I wanted to hear nothing of it, because I was dreaming of bi-focal glasses at the time.[38]

---

[36] Poznań International Fair – the biggest industrial fair in Poland which was established in 1921.

[37] Krzysztof Matlak, "Sława, pieniądze i…," *Głos Szczeciński*, July 18–19, 1987: 6.

[38] Anna Dulemba, "Skok na antenę," *Polityka*, supplement: *Polityka-Eksport-Import*, August, 1986: 1.

Anteny satelitarne paraboliczne

Dane techniczne:

1. Rozmiar	(mtr.)	0,90	1,20	1,50	1,80
2. Ogniskowa	(mtr.)	0,36	0,48	0,60	0,72
3. Ogniskowa /sredn.	(mtr.)	0,40	0,40	0,40	0,40
4. Uzysk 11,3 GHz	(dB)	39,0	41,5	43,2	44,6
5. Uzysk 12,1 GHz	(dB)	39,6	42,1	43,9	45,2
6. Waga	(kg)	8,5	13,5	28,0	38,0
7. Max.sred.masztu	(mm)	50	60	70	110

8.Lustra wykonane z blach aluminiowych o grubosci 1,5 do 2,5 mm

9.Mocowanie luster typu biegunowego   stalowo   cynkowane galwaniczne.

Producent:

Zakład Slusarski

Adolf Bogacki

ul. Gen. Dabrowskiego 11

tel. 80-440   tlx 422739

70-100 Szczecin

Jednostak Innowacyjno - Wdrozeniowa

Rozpoczecie produkcji anten: 1986 r.

Cztery zastrzezenia rozwiazań konstrukcyjnych zamocowania

zarejestrowane w Urzedzie Patentowym RP.

Ponad 60% produkcji jest sprzedawana na rynki zachodnie.

Fig. 4. An advertisement for the Adolf Bogacki metal workshop

Source: Adolf Bogacki's private archive

And yet, despite this initial reluctance towards another change of path in his career, Żniniewicz decided to take the risk. He began encouraging Adolf Bogacki, who, like him, had a workshop in Szczecin's suburbs, to collaborate:

So he says to me: "Starting tomorrow, we're working on satellite dishes." I reply, "What are you, insane? I've no idea how to, nor…" But [he] says: "I've got the people to do the maths, all we have to do is make it and put it together." And that's how it was.[39]

The results of this crazy project outgrew everyone's expectations. Here is how Żniniewicz himself described the first "Made in Poland" satellite dish test in 1986:

> We're setting it up, we're looking for the satellite signal and suddenly, wow! We're receiving one channel, a second channel, then a third. These Swedes I was working with never drank vodka, but this time all of them were completely smashed, myself included. Both they and I knew what was coming: we had millions of Kronor on its way to us.[40]

The alcohol-driven enthusiasm shown by the Swedes was entirely understandable. The production of satellite dishes in Poland had one major advantage. As Bogacki succinctly put it: "It was cheaper. Cheaper, cheaper, and once again, cheaper."[41]

Bogacki also lists the main reasons why running a satellite dish business from Poland was an incredibly profitable project. Besides the labour cost, the instruments themselves were also less expensive to make in Poland. The exchange rate of the US dollar in the PPR was so high that the actual price of the dishes in the West was several times higher than what the domestic manufacturer received.

One thing worth clarifying is that, initially, the craftsmen from Szczecin were not producing whole satellite dishes, but only the reflectors. As Bogacki points out, all the rest – like the zinc plating in the rear part, and painting in powder-coating paint shops – took place in Sweden.[42] There is, however, no doubt that the creation of the reflector was the most demanding part of the process. It required designing the form on the one hand, and the creation of adequate production technology on the other. Żniniewicz guarded his secrets very carefully, especially those pertaining to the method of form production, because, like he said: "there were people from huge companies coming here to find out how we were doing it."[43]

---

[39] Adolf Bogacki, interview, September 15, 2015.

[40] Dulemba, "Skok."

[41] Adolf Bogacki, interview, September 15, 2015.

[42] What is interesting is that there were no powder paint shops in Poland, so the antennas sold on the domestic market were not painted at all. A natural process of aluminium oxidation made the antenna matt enough for it not to reflect sunbeams which could damage the converter.

[43] Dulemba, "Skok."

Bogacki was less secretive, jokingly adding that if someone had wanted to use the Szczecin technology, they would have gotten his permission. As it turned out, the making of an instrument as precise as a satellite dish, which receives signals coming down from space from a height of nearly 36 thousand kilometers, was possible with what were, essentially, do-it-yourself methods:

> [...] the mould for the mirrors to be pressed in was made of concrete. Yes, seriously (*laughter*). Making a parabolic reflector is not all that difficult either, but we didn't build presses. [The mirrors] were simply compressed by air (*laughter*). This was our idea, so to test it out first we made a small antenna. We figured that if the small one came out alright then the big one would as well. All we needed was more air (*laughter*). I think I have one of these concrete moulds somewhere here still.[44]

Putting it in the simplest way possible: there was a metal construction into which a piece of aluminium was placed. When the aluminium was attached with screws, compressed air was released through a special opening. As a result of the pressure it exerted on the plate, the aluminium took on its parabolic shape. Of course, this not-so-sophisticated production technology of the Szczecin antennas had numerous weaknesses, of which the concrete mould was the greatest. But, according to Bogacki, even that was dealt with: "If a hole appeared, we would add more concrete (*laughter*). It was as the bricklayers say: pour water, pour sand, and do not skimp on material."[45]

Fig. 5. Zdzisław Żniniewicz with his travelling satellite television in Warsaw

Source: *Ekran* 24, 1987, 14

---

[44] Adolf Bogacki, interview, September 15, 2015.
[45] Ibid.

It is, of course, hard to believe that the antennas which were produced this way were in any way of equal quality to the devices which were made in the West. Besides, Bogacki himself openly lists the main weaknesses of his product:

> [...] I have to say that the Swedish dish was made of sheet metal which was at least 60–70 percent thicker. Those plates were stiff, and ours were supple in comparison. Their antennas were pressed in their entirety, while we pressed the reflectors. But the back was produced in a completely different way.[46]

The Swedes who sold these dishes were also very aware of the weaknesses, which is why they would go to great lengths to conceal the fact that they came from Poland. When the Swedish press published an article about Żniniewicz and his product, the Swedes panicked. This is how Bogacki recounts it:

> Over in Malmö, in their newspaper, a two-page article was printed about a Polish satellite dish manufacturer from Szczecin... And at the time – it is different now – they considered us to be the wilderness. They thought it was all bears, not satellite dishes. Anyway, the contractors warned him that should any more dishes from Poland show up over there, they're cancelling the contract. You see, when someone bought a dish in Sweden, they assumed that this dish is a Swedish product, not some Polish one.[47]

All these construction imperfections aside, Szczecin's satellite dishes were great in two ways. First, they worked, and second, they were made with aluminium, a "noble" material. What the interview with Marek Czajkowski proves is that not all domestic satellite dish manufacturers who followed in Bogacki and Żniniewicz's footsteps made their dishes from "sheet metal." For example, his Warsaw suburb supplier used epoxy resins to press the reflectors:

> We had this manufacturer who made boats out of plastic, these canoes. We brought a satellite dish from outside the country and commissioned him to make a mold for one just like it. Based on that one dish, he manufactured the dishes for the receivers that we were bringing in from Sweden. As it turns out, he ended up making quite a lot of these dishes. He even stopped making the boats, because he was making dishes for us all the time.[48]

I have mentioned the material which the satellite dishes were made from, but not only to denote the value of the craftsmen's product. We must remember that in the 80s, the private sector had very limited access to raw materials. As a result,

---

[46] Ibid.
[47] Ibid.
[48] Marek Czajkowski, interview, September 16, 2015.

acquiring aluminium was no less of a challenge than building the mould or thinking up the production technology. At the initial time of their project, Bogacki and Żniniewicz had to turn to "black market" methods of acquiring the scarce material. Once the Polish satellite dishes became a big deal, however, the communist authorities began looking at the craftsmen from Szczecin more favourably. Adolf Bogacki's workshop gained the status of a research and development unit, which resulted in measurable benefits: "When I had that research and development unit, I could buy machines, I could erect buildings, and certain exemptions applied to me (investment exemptions), at least, when I paid taxes."[49]

Bogacki does not hide his feelings on the ways the government supported private enterprise in the past:

> Everyone believes that everything during the days of the so-called "komuna"[50] was bad, but it wasn't that way at all. You knew there were intelligent people about, because we had these research and development units; there was a government bill passed for those.
>
> And, work was being done, people were employed. In the case of export, foreign currency came in. The government's attitude was, why not help? We won't take taxes if he builds something.[51]

It is possible to see such economic initiatives, which took place towards the end of the PPR, as effective. Likewise, Bogacki's recollections of research and development units seem to cast a positive light on the welfare state attempting to regulate the market as much as it could. But Dariusz T. Grala states:

> The regime in the 80s attempted to make the government sector more effective by implementing a project of economic reform [...]. Meanwhile, it was not the state-owned establishments but private businesses that showed a high speed of growth when it came to the number of enterprises, employment, and sales. This situation did not appeal to many upper-tier members of PZPR (Polish United Worker's Party). Due to this, economic policy included many barriers aimed at limiting the growth of the non-socialised sector.[52]

It is also worth reminding that towards the end of the 80s, the press began to criticise research and development units. In her article "Pod górkę" ("Uphill"), Magda Sowińska points out that

> two years after the innovative unit act was passed, it stopped having a reason to exist. Indeed, since the beginning of the year, representatives of the units have been wondering what preferential treatment they are supposed to be getting. New solutions

---

[49] Adolf Bogacki, interview, September 15, 2015.
[50] Contemptuous and colloquial term for the communist regime in Poland.
[51] Ibid.
[52] Grala, *Reformy gospodarcze*, 268.

which now function in our economy, especially those which began in January 1989, guarantee more beneficial conditions to all enterprises, and at least are comparable to those which are included in the research unit act.[53]

It seems more likely that the success of the Antenna Guys from Szczecin, similarly to the importers before, was aided by a certain loophole in the planned economic system. In this case it was not only about delivering a product, which was out of reach for state-run enterprises, but about processing raw aluminium and selling it in the form of a satellite dish, for a higher price. As proven by numerous press articles from the end of the 80s, domestic export was indeed pathological. The export market was led exclusively by institutions which dealt with the extraction of raw materials (e.g. coal and sulphur mines) and possibly their initial processing (smelters). However, no one was paying attention to the significance of the development of businesses which processed raw material into specific products. On the one hand, this type of export strategy guaranteed direct profit without having to bear the cost of investment in the technologically-delayed industry. On the other hand, due to this economic short-sightedness, the state made enormous losses because the raw material is always several times cheaper than the product made from it.[54] The Antenna Guys from Szczecin were clearly going against the mainstream of the PPR's inefficient export strategy, which contributed to their success.

Finally, it is worth adding that Żniniewicz had incredible marketing skills, which he used equally well inside and outside of the country. In an account from a reporter in the previously mentioned *Głos Szczeciński*, there is a reference to the uproar that Żniniewicz's satellite dishes made at the Poznań Fair. According to Bogacki, there were crowds pushing towards the satellite dish stand:

> I think it was 1986, the Poznań Fair. […] the most crowded stand of all was the satellite television section, which was in a trailer outside. I remember my wife and I were coming home from vacation and I said, "come on, let's drop by the fair." When we got there I thought, gosh… All I could see was young people glued to these TVs. From what I recall, there were two or three [TVs], playing a constant stream of music from the West. We wanted to go nearer to be able to see better, but there were so many young people there's no chance we could have been able to.[55]

Besides, it was not just Poznań that got to taste the charms of satellite television. In that same year Żniniewicz and his travelling SAT TV stand made their way to Warsaw.[56] This is how Marek Czajkowski recalls the event: "I was there at

---

[53] Magda Sowińska, "Pod górkę," *TOP*, October 13, 1989: 13.

[54] See Jerzy Baczyński, "Przeciąganie kabla," *Polityka*, supplement: *Polityka– Eksport – Import*, April, 1986: 4.

[55] Adolf Bogacki, interview, September 15, 2015.

[56] Franciszek Skwierawski, "Telewizja satelitarna ruszyła…," *Ekran*, June 18, 1987: 14–15.

the showcase in front of the Palace of Culture and Science, where he was telling people how it all worked. Anyone could come up and see."[57]

Without a doubt, these innovative (for their time) marketing campaigns contributed to the spreading of awareness about the phenomenon of satellite television among Polish people, and as such, they lay the foundations for the future satellite dish market. This market grew very quickly, as evident from Adolf Bogacki's words: "There was a massive demand for them in Poland, Ukraine, Belarus, Germany."[58] Domestically, the most important clients that the Antenna Guys from Szczecin had were fitters such as Jerzy Lubacz, who was supplied by Bogacki. Of course, with such an absorbent market, new satellite dish manufacturers appeared. As Bogacki claims, at the beginning of the 90s, many dishes appeared in the Podkarpackie province and in Warsaw.

As in the case of the importers of electronic components of SAT TV sets, the satellite dish manufacturing sector was fully formed at the dawn of the Third Polish Republic. All that was left to do was pressing the sheet metal and implementing further improvements. For example, over the course of a few years Adolf Bogacki patented many new solutions in the area of dish attachment and acquired his own powder coating facility around 1995.

Cable television to the best of our possibilities

The book *A Guest or an Intruder from Space?*, which offers an in-depth and definitely critical analysis of the phenomenon of direct-reception satellite television, also features a proposal of an alternative solution to the problem of access to audiovisual programmes streaming from space.

The second possibility is to install a system of receiver stations from which the programmes are then directed to the individual user with a cable network. In this case, the majority of the cost is taken on by the state. These costs are retrievable after a certain amount of time through radio and television fees though. However limited the possibilities of forming individualised programme reception are in this system, it is more democratic because it allows access to satellite television for a greater number of families – even those of average wealth. Besides, a cable network can be used not only for satellite television, but also for various telecommunications. It constitutes an effective security measure against diversionary programmes.[59]

---

[57] Marek Czajkowski, interview, September 16, 2015.
[58] Adolf Bogacki, interview, September 15, 2015.
[59] Fryszkiewicz, Grabski, Sarosiek, *Gość czy intruz*, 14–15.

Even though this paragraph provides several arguments for the collective reception of satellite television, the last one was, without a doubt, most important from the authorities' point of view. A cable network would provide the comfort of full supervision over the content that was being broadcast, much like SAT TV retransmissions on state-run television stations. In other words, the socialist CATV offer could be narrowed down to include only those broadcasters who presented content that was ideologically aligned with the state.

Fig. 6. The first day of the Ursynów experiment described in the local press.
Headline in blue reads: "There is image but no sound. There is sound but no colour."

Source: *Pasmo*, February 27, 1988: 1

This incredibly consistent and, at the same time, promising plan to tame the television from space had just one flaw. This flaw, as the authors of the book *A Guest or an Intruder...* point out, is the fact that "the majority of the cost is taken on by the state."[60] But, in the second half of the 80s, the state was incapable of undertaking any constructive steps towards the development of a media infrastructure. This is confirmed by the following excerpt from Jacek Mojkowski's article published in *Polityka* in March 1987:

> [...] the postulate is as follows: **to develop cable television in Poland** [original emphasis] commonly used in many Western countries. For those who are unfamiliar, let's just say that TVs in Western countries do not need dishes to receive programmes, because they are delivered via cables. This is why there are no problems with the placement of parabolic antennas in the West – they are not necessary. Signals from the satellite are caught by one central antenna, and immediately sent further down cables. This is, therefore, a kind of megaphone with many channels, but thanks to this system it is less costly and easier to operate than the traditional one. With that in mind, we must ask: how do we pay for it and what materials do we use to make it? The state expenditure on television is half of what it was in 1980. We are short on telephone cables, so why even dream of cable television? Many branches of the electronics industry are being made to go back to manufacturing, however, there is no way of being sure that they will cope with the production of devices which would receive satellite television.[61]

The technological backwardness of socialist Poland described here suggests that cable television in the PPR was pure fantasy. The author did not consider one important factor in his argument, however: the significance of grass-roots social initiatives. As proven by the neighbourhood television in Ursynów,[62] even in situations where there was an extreme lack of equipment and cables for the distribution of audiovisual signals, with a bit of good will and the involvement of qualified people it was possible to create a collective satellite television reception network in the PPR.

It all began in January 1988, within the pages of *Pasmo*, a periodical published by the Ursynów-Natolin Association for Society and Culture. An interesting advertisement appealed to its readers:

> A thing bordering on fantasy, yet true: satellite television in every flat for 500 złoty per month. No walls need be torn down, no antennas installed on the rooftops and no costs amounting to millions. Here are the technical details:

---

60   Ibid.
61   Mojkowski, "Świat," 16.
62   Ursynów is a district of Warsaw.

Fig. 7. A satirical drawing from the *Pasmo* magazine. Satellite television was installed in Ursynów before phone lines

Source: *Pasmo*, March 12, 1988: 1, 3

In chosen buildings where all residents give their consent, through the use of an existing installation of a collective antenna (sockets, wiring), modulators will be installed. These modulators will function at a bandwidth of TV channels 1–12; four satellite television channels will be added, along with two public television channels and one Soviet channel. Which satellite television programmes will be chosen is up to the residents to decide, and the option of changing them at any given moment remains. One receiver can feed the entire enclave of 12 buildings. All of the equipment is imported, so the modulators in the buildings and the satellite receiver system will be delivered by the Swedish company PORION AB, and their owner Marek Czajkowski […]. In short, this is the possibility that is opening before us thanks to the initiative of the Ursynów-Natolin Association for Society and Culture. PORION AB can wire the entire residential district within a month […].

Please send your applications. We also ask the house committees to get the permission of all of the buildings' residents, because PORION AB needs their permission to take on the entire project.[63]

The housing estate's residents' response exceeded all expectations. This is what a columnist for *Pasmo*, Marek Kasz, wrote about the topic three weeks later:

Even though it is not here yet, satellite television has already brought about positive changes. First and foremost, it has compelled many thus-far apathetic residents to band together and start collecting signatures. These are people who, like me, could not be convinced to join community projects to convert waste land into parks and tennis courts. Now, they are quite happy to be harassing their caretakers to compile lists of residents wanting to install satellite television, and to make them do it on time.

Without a doubt, the main reason for this common excitement was the prospect of gaining access to foreign channels for the symbolic sum of 500 złoty (at the same time high-end satellite television reception set could cost up to 8 million złoty). However, the cost reduction came with certain limitations, which the owner of PORION AB Marek Czajkowski told residents in advance:

Let us make this clear: this will not be cable television in the typical meaning of the word. Cable television requires a separate, independent network which would need to be directed to the apartment of any person wanting to use it. A television centre could theoretically retransmit all satellite programmes which reach Poland and even more – if it had a studio, it could broadcast its own informative programmes, VHS films, etc. […] Our offer is more modest. We propose a dozen separate installations on the housing estate, each with its own satellite dish and equipment necessary for the processing of the satellite signal. This signal, when properly modulated, would be carried to apartments with the already existing installation of a collective receiver.[64]

Even though the installation proposed by Czajkowski bore all the marks of poor people's television, the members of the Ursynów-Natolin Association, as well as the housing estate's residents, were keen to go ahead. As the editorial office of *Pasmo* stated on February 6:

---

[63] Taw., "Wielki kabel na Ursynowie," *Pasmo*, January 9, 1988: 1, 3.

[64] Andrzej Gorzym, "Jestem dobrej myśli… Rozmowa z Markiem Czajkowskim, właścicielem szwedzkiej firmy PORION AB," *Pasmo*, January 23, 1988: 4.

Up to now, we have received over 100 lists of signatures from various housing estates in Ursynów and Natolin. Some committees have already gotten the permission of all the residents in their area [...]. There are few people against or undecided. Nowhere do their numbers exceed 10 percent.[65]

The article concludes with an observation that is somewhat far-fetched in its political expression:

This mass support for the Ursynów-Natolin initiative shown by the Ursynów-Natolin Association for Society and Culture is undeniable proof that a collective drive for satellite television has come. The residents' attitude shows that a time for social activism has also arrived.[66]

The communist authorities must have had mixed feelings about these kinds of declarations. On the one hand, the prospect of creating a collective satellite television reception network from the bottom up (and, additionally, from scratch), which could later be copied in subsequent cities, was unusually attractive. Supervising television from space as proposed by the authors of the book *A Guest or an Intruder...* suddenly seemed realistic. On the other hand, community enthusiasm and the strength of a grass-roots initiative which accompanied the Ursynów-Natolin experiment had to give rise to valid apprehension.

The decision of the National Radio Inspection Office, which allowed for the experiment to take place during a time when Poland had no laws regulating collective reception of satellite television, shows that the authorities took an approach of limited trust towards the Ursynów-Natolin initiative. According to the initial agreement, the Association received a due date of May 15. If, by that time, the Porion company could manage to launch the installation, the following years would see the rollout of an ambitious broadcasting system, encompassing the whole Ursynów-Natolin estate. At the same time, the authorities were clearly letting the community know that at any given moment they could put a stop to the whole thing, because there was too much uncertainty hovering over the initiative. This is what was written in a press-published account from a meeting of the members of the association with representatives from "just about all offices and institutions dealing with television":

The concerns pertained mostly to the legal aspect. In essence, this entire Ursynów-Natolin experiment comes down to illegal viewing of someone else's programmes. To observe the international norms, international agreements with

---

[65] SAM, "Telewizja satelitarna: strzał w dziesiątkę," *Pasmo*, February 6, 1988: 1.
[66] Ibid., 2.

satellite owners and, separately, programme owners would have to be made, and appropriate fees paid. Only then could programmes be received in accordance with the law.[67]

It is easy to see how the communist authorities are, with calculated hypocrisy, repeating arguments that two years prior were supposed to prove the illegal character of individual satellite TV recipients' activity. However, as we read towards the end of the press report:

The debaters have mostly agreed that our experiment (although it is, in essence, satellite TV for the poor) is the cheapest – and therefore, in the country's current economic situation, the only one possible in the next decade, that offers a solution meeting social expectations.[68]

Porion quickly achieved what it set out to do:

On Tuesday February 16, the anticipated moment took place: reception of the experimental satellite TV programme at Residents' Club in Imielin, at Marco Polo Street 1, had begun. At 5 pm the entrance was packed. The TV at the club was surrounded, transmitting SAT 1 and SUPER CHANNEL channels, received so far by residents of buildings at streets Marco Polo 1 and Miklaszewskiego 3.[69]

It is evident from the article that the Ursynów-Natolin experiment would soon be replicated across the country:

Our attempts have piqued the interest of residents of other cities. We've had representatives form Olsztyn, Poznań and Opole get in touch. Two activists from Kwidzyń have reported that the local Social Society wants to take the risk of running a similar experiment.[70]

Despite these far-fetched declarations, during the next few months the Imielin estate invariably remained the only battleground for cosmic television for the masses:

Attempts have been made continuously throughout the whole of February and are still being made. Initially, only one channel was introduced. Then a second one. It was immediately evident that only the owners of TVs adapted to the PAL system will be able to receive them. Therefore, a way of transforming to the SECAM

---

[67] SAM, "Wśród serdecznych przyjaciół," *Pasmo*, March 5, 1988: 2.

[68] Ibid.

[69] SAM, "Telewizja satelitarna w próbach," *Pasmo*, February 27, 1988: 1.

[70] Ibid., 2.

system had to be found. Simultaneously, attempts were made to introduce a third and fourth channel to the network. The fourth one isn't working. There are hopes that eventually it will, but there is still no certainty.[71]

However, the turning point of the entire experiment came at the beginning of the second month of its operation:

On March 5, at 7 pm exactly, the first ever half-an-hour-long local TV programme in Poland was transmitted. So far it's been called experimental, but it had a fully professional character. It was hosted by a top-class presenter: Edyta Wojtczak, with Andrzej Ibis Wróblewski co-presenting.[72]

This event was important for at least two reasons. First, it illustrated that neighbourhood television, hastily put together by a group of enthusiasts, could serve communication goals other than just the passive distribution of satellite TV channels. Second, it gave the entire experiment a new, political dimension, especially when one considers the fact that all the following local programmes (eight were emitted in total) pertained to important issues of the Ursynów community's life, and representatives of the cooperative, police and city council were invited as guests. Moreover, just before the National Councils' elections[73] on June 14, a kind of election night special was organised at the URSYNAT TV studio, from where the programme was filmed. During the transmission: "The candidates faced such a barrage of questions and complaints, that they had to work really hard to save face."[74]

This pro-social aspect of Ursynów's local television was not a product of chance, but stemmed from its creators' deeper beliefs. Wacław Tylawski's following statement illustrates this perfectly:

[...] integration is possible when people come into contact with each other. Where are we to seek such contact, when there is no promenade, no park, not even a coffee shop? The only available solution, brought to us by technology, is cable television. People can introduce themselves to each other, and they can talk – through a camera and a cable for now. Maybe, however, these bonds will get stronger, maybe people won't be strangers to each other anymore ... Perhaps they'll become neighbours and feel a sense of responsibility for our little homeland.[75]

---

[71] IBIS, "PIR robi pomiary," *Pasmo*, March 12, 1988: 1.

[72] Radosław Zięcina, "Tu studio URSYNAT," *Pasmo*, March 19, 1988: 1.

[73] The National Councils were the local organs of state power. They were supervised by The State Council.

[74] SAM, "Wiadomości z frontu TV-SAT," *Pasmo*, June 25, 1988: 1.

[75] SAM, "Twórca TV-SAT na Ursynowie. Rozmowa z inż. Wacławem Tylawskim," *Pasmo*, June 11, 1988: 4.

It is worth noting here that, without Tylawski, Ursynów's local television would never have happened. This is because he, as a high ranking employee of Polish Television, was the only person who could "borrow" the equipment necessary for making the programme (of course, the entire operation was completely illegal and based on a network of informal connections, which was typical for the reality of the PPR). Several colourful anecdotes can be found in *Pasmo*, relating how this borrowed equipment served Ursynów's local television pioneers:

> Marek Jefremienko was the camera operator, who, in spite of receiving several electric shocks thanks to certain equipment malfunctions, would excellently frame the picture and steady his hands after. All the cameras, microphones, Betamaxes and other marvels of technology were connected by Wacław Tylawski, who elegantly edited a report on failures in local waste collection. The studio audience was galvanised by what they saw and are looking to form a cooperative manufacturing rubbish bins.
> Like with any television broadcast, there were some technical difficulties. Whether because the voltage was too high, or because of a short-circuit caused by Jefremienko's body, the power went out. It wasn't possible to show the "Please Stand By" board, as the studio hadn't acquired one yet. After the power came back on, however, the residents could watch Jolanta Pieńkowska, hosting the programme with great poise.[76]

Regardless of how difficult and complicated the birth of the first local television in Poland was, it was met with enthusiastic reception from the local community. The evidence of this is an extraordinary situation that took place during the transmission of the fourth URSYNAT studio broadcast:

> Something happened that made us all at URSYNAT Studio extremely happy. The engineer Wacław Tylawski called on the viewers to go to their balconies and wave at the cameraman standing in front of the building. Instantly, the balconies filled with people. It was beautiful! On a beautiful Saturday, when people usually go away for the weekend, we weren't speaking into a vacuum.[77]

Unfortunately, this promising local initiative, which in the Third Polish Republic could have been utilised towards building a mature civil society, came to naught. Sharply rising inflation made the sum of 500 złoty, cited as the cost of cable at the beginning of the experiment, insufficient to cover the costs

---

[76] Radosław Zięcina, "Mimo braku zasilania studio URSYNAT nadaje," *Pasmo*, May 21, 1988: 4.
[77] SAM, "'URSYNAT'" zwycieża!," *Pasmo*, May 28, 1988: 2.

of network operation. Although the National Radio Inspection had agreed to extend the phase of technical trials until the end of 1988, in the following year the residents of the Ursynów-Natolin estate lacked the will to expand the network and turn it into a professional cable television network.

The satellite promised land

Although the Ursynów-Natolin Association for Society and Culture's initiative never got past the experimental phase, it had a profound influence on the development of neighbourhood television across the country. First of all, it was because of the actions of Warsaw's satellite TV enthusiasts that the Ministry of Transportation, Maritime Affairs and Communications decided to legalise collective satellite TV reception networks, through a special decree. From November 1 1988,[78] it was not just people that owned receiver sets for private use that could apply for a satellite TV access permit, but also bottom-up social committees aiming for mass scale satellite signal distribution (though they could only apply to the National Radio Inspection through legal entities: that is, housing cooperatives or workplaces).[79] Secondly, thanks to the Ursynów experiment, Porion began to specialise in importing equipment for collective satellite TV reception, which was sold – and, if necessary, installed – in different Polish cities by its authorised representatives. As a result, during the following months, Marek Czajkowski was directly or indirectly handling satellite initiatives throughout the country. There were many such initiatives, as best evidenced by the following excerpt from the magazine *TV-Sat-Magazyn*:

> Satellite television in Poland, especially in the form of neighbourhood television, is developing more and more rapidly. Undeterred by numerous unfavourable regulations and limited technical abilities of the equipment in use, multiple social committees spring up like mushrooms in many cities and housing estates. Their first aim is to figure out the number of those potentially interested in watching programmes from space, then to acquire necessary funds and finally, to hire a specialised company for the entire installation.[80]

---

[78] Sylweriusz Ładyński, "Jak zorganizować własną telewizję osiedlową (6)," *TV-Sat-Magazyn*, October, 1989: 29. It is worth noting that applications for TV SAT reception permits were submitted to the National Radio Inspection after the local television had been installed. For the permit one had to submit a request via a special form printed by the post office and technical documentation of the TV SAT cable television.

[79] Machejek, "Odmówić."

[80] A. Marciniak, "Satelitarne vademecum," *TV-Sat-Magazyn*, April, 1989: 2–3.

Although the trend for neighbourhood television became commonplace and, like in Ursynów, unleashed limitless amounts of social enthusiasm in millions of Poles, the phenomenon of satellite TV for the poor quickly found its critics. Franciszek Skwierawski, the biggest promoter of satellite TV in Poland, turned out to be one of them. As he wrote in *Rzeczpospolita* in 1991:

[...] most big housing estate residents are condemned to collective reception. They have few solutions to choose from.

The most primitive and cheapest is to adapt an existing collective TV antenna to receive three to four satellite channels, which doesn't guarantee the correct technical quality and leaves out the residents not participating financially in the endeavour. This kind of installation is commonly referred to as cable TV for the poor. Companies - that's too generous a description, people, rather - offering such an "investment" are getting the money, doing the work quickly and running away, covering their tracks to avoid dealing with complaints, because it often turns out that the modifications result in problems with the reception of Polish channels. This results in those not signing up holding grievances against the initiators of the entire enterprise.

A slightly better form is a satellite TV cable installation that is independent from the shared antenna, covering only the apartments whose owners had signed up for the project. There are many companies offering such services in Poland, a classic example being the "Video-SAT" from Dzierżoniów, whose owner has convinced residents from over 30 cities in Poland to use his services. These companies are offering pseudo-cable TV, in other words, a primitive installation, allowing for distribution of several foreign channels. The problem, however, lies in the fact that the cables running between buildings are usually hung on streetlights and supplied to the apartments via external walls. Many of these companies hold licenses issued by the communications department and documents stating that their equipment is domestically compatible.[81]

Apart from the purely technical shortcomings of neighbourhood television, Skwierawski pointed out one additional problem with collective satellite TV reception in transformation era Poland:

We must face the truth. In our country, the sale of somebody else's property in the form of foreign TV programmes is taking place on an ever wider scale. So far, no one has made an effort to address this situation, questionable from a copyright viewpoint. Domestic cable network owners should get permission from satellite channel owners to distribute their programmes, like everywhere else across the world [...] The scale of the phenomenon that is cable TV in Poland, whether

---

[81] Franciszek Skwierawski, "Kowalski i komplikacje satelitarne," *Rzeczpospolita*, August 31 – September 1, 1991: 8.

of amateur or professional character, is starting to gain momentum. It must be taken into account that foreign satellite channel owners will stand up for their rights [...].[82]

As far as the last of the objections articulated by Skwierawski goes, it has to be noted that he is far removed from the dishonest statements made by communist authorities, who, in the 80s, invoked copyright laws only in order to halt the development of satellite TV in the PPR. With the economic and political transformations that took place in Poland after 1989, the problem of copyright protection of audiovisual content created by foreign producers ceased to be purely abstract. The Polish government, as well as telecommunications experts, quickly realised that if the country was to begin its journey towards democratisation and consequently partake in the cultural benefits of globalisation, it was necessary to develop legal regulations and best practice regarding satellite TV access.

On the other hand, we must consider potential bias in Skwierawski's objections towards the phenomenon of neighbourhood television itself. Admittedly, many dishonest satellite TV installers appeared in Poland during the 80s and 90s, such as the company from Dzierżoniów mentioned by the *Rzeczpospolita* journalist, whose owner had failed to deal with a large project in Warsaw. But in many other cities where satellite television for the poor was the only option available, the experiments following Ursynów's initiative resulted in complete success. This was especially true of Łódź, a city where, in the early 90s, several neighbourhood televisions were operating.

The first installation for collective satellite TV reception appeared in Łódź during the summer of 1988, only a few months after Warsaw. Residents of the first four buildings started receiving programmes from space in July, and at the beginning of September the signal reached the remaining residents of the "Piastów" estate. Here is a short description of the entire project published in *TV-Sat-Magazyn*:

> In total, the satellite TV cable network created covered forty buildings (51 collective antennas) and 2,155 apartments. 4.5 km of concentric cable was installed, resulting in the biggest neighbourhood satellite TV network in the country. It was implemented during a period of no regulations governing the issuing of collective satellite TV reception permits and there were no domestic models to follow. Those were the two basic factors that strongly hindered the project's credibility in the eyes of the residents from Piastów estate, before it was, in effect, proven that the "impossible" is possible after all.

---

[82] Ibid.

It all happened thanks to the efficient organisation and generosity of several buildings' representatives, the benevolence of "Czerwony Rynek" Housing Cooperative's management, and the majority of the estate's residents, for whom the cost of installation wasn't too high, as well as because of the affordability of the single apartment fee.[83]

More specific information about the preparations for the launch of the collective satellite TV reception system on Łódź's "Piastów" estate, and later, the installation works, can be found in a series of articles under the title "Jak zorganizować osiedlową telewizję satelitarną ("How to Organise Your Own Neighbourhood Television"), published in *TV-Sat-Magazyn* and authored by the project's initiator, Sylweriusz Ładyński.

The entire project started with A4-sized notices posted in every staircase, with the following content:

Satellite television
In June of the current year provisions will be made for the possibility of connecting the "Piastów" estate's buildings to a satellite TV network. The reception of satellite TV programmes will take place free of charge, around the clock, on every kind of TV without the need for modifications, but using the existing collective installation.
The expected cost per apartment will be around …. one thousand złoty and will be dependent on the number of people interested.
All the residents will be informed about the method and time of payment.
Representatives of the individual buildings willing to help with the organisation within their building are welcome.
Contact at the number … … ….[84]

"Piastów" estate residents reacted immediately. During the three days that followed the posting of notices, Sylweriusz Ładyński's phone was ringing constantly. The people who were first to contact the project leaders for information about the free, publicly available satellite TV were offered involvement in the organisation. Since most of them agreed, a group of around a hundred representatives (one person per staircase) was quickly assembled. After these initial preparations came time for specific action.

With help of the assembled team of building representatives, residents of the "Piastów" estate were given personal declarations to complete. A one-day deadline was set for signing them. Next, the declarations were collected and delivered to the Main Organising Committee of the neighbourhood satellite television. After

---

[83] Sylweriusz Ładyński, "Łódzkie doświadczenia," *TV-Sat-Magazyn*, May, 1989: 25.
[84] Idem, "Jak zorganizować osiedlową telewizję satelitarną (3)," *TV-Sat-Magazyn*, July, 1989: 26.

the declarations were calculated, it turned out that as many as 80% of the residents had committed financial contribution to buying satellite TV reception equipment. Out of this 80%, 55% of residents committed contributions of between 20 and 150 thousand złoty, and 45% up to 10 thousand złoty. [...] In the end, the contribution was set at 15 thousand złoty and 10 USD for people declaring foreign currency.[85]

For the most part, the residents' positive reaction was resulting from the information campaign, skillfully and meticulously conducted by the "Main Organising Committee":

During the distribution of declarations to all the residents of the "Piastów" estate, it became apparent that only a small group knew what satellite television was. The rest had heard about it, but there were also residents who had never heard the term before. In order to advertise the benefits of satellite TV and increase the number of those interested in financing the project, leaflets with information about satellite television were posted in the staircases: leaflets that went missing immediately. It has to be presumed that some of the residents, hungry for knowledge on the subject, were reading them at home, while others, consequently, remained uninformed. A simple idea of how to inform the future subscribers about satellite television goes as follows: the essential information on the subject must be printed on the reverse of the declaration in A4 format. Before the future satellite TV subscriber fills out and signs the declaration, he'll familiarise himself with the description printed on the reverse. This strategy works every time, without a miss!

When it came to it, collecting the payments was not without its problems. Only 75% of declared contributions were paid in the time set by the organising committee, and, among the 333 people who declared contributions in American dollars, less than half decided to pay in this way in the end. Fortunately, with the installation works advancing, the unconvinced 25% gained trust in the venture and paid the appropriate charges.[86] This made it possible to finish the project in a short amount of time.

New problems, however, arose with the launch of the satellite TV network. Since the AZART installation theoretically allowed for distribution of no more than three channels (and practically one), from the very beginning a sense of dissatisfaction rose among the recipients, who were aware that far more channels were flowing to earth from the sky. It was reflected in numerous requests, which were submitted mostly by telephone, for the schedule change of channels distributed in the satellite TV network. Members of the organising committee immediately concluded that demands made by disgruntled residents could not be met, as it would lead the entire operation to the brink of disaster:

---

[85]   Ibid.
[86]   Ibid.

For the numerous and varied group of collective satellite TV subscribers, vastly different programmes are interesting at any given time. A person enquiring about switching the currently received channel for a different one must understand that their choice might not align with the taste of the others. This was verified experimentally on the "Piastów" estate. A subscriber enquiring about changing the received channel was asked to leave her name and phone number before the channel was changed accordingly to her wish. People who reacted to the change with telephone complaints were directed to her. Several minutes later, the same subscriber called again asking that the channel be switched back to the previous one.[87]

It was of course, inconceivable for the members of the organising committee to educate every dissatisfied subscriber in this manner. Therefore, a different solution was adopted, which intended to neutralise the satellite discontent among the residents:

It was agreed at the "Piastów" estate that a different channel would be transmitted every day, because unjustified changes throughout the day resulted in suspicions that the changes were about censorship and good programmes would be missed. The accepted schedule for the "Piastów" estate was biweekly and presented as follows:

Week I (vertical polarisation)
Monday – 6.00 – SAT 1
Tuesday – 1.00 – SUPER
Wednesday – 6.00 – SAT 1, 17.20 3 SAT
Thursday – 6.00 – SAT 1, 17.30 TELECLUB
Friday – 1.00 – SUPER
Saturday – 7.00 – SUPER, 15.30 TELECLUB
Sunday – 7.00 – SUPER, 14.40 RTL

Week II (horizontal polarisation)
Monday – 1.30 – SKY
Tuesday – 6.00 – RTL
Wednesday – 1.30 – SKY, 16.05 TV 5
Thursday – 1.30 – SKY
Friday – 6.30 – RTL
Saturday – 1.30 – SKY, 18.00 RTL
Sunday – 8.00 – RTL[88]

---

[87] Sylweriusz Ładyński, "Jak zorganizować osiedlową telewizję satelitarną (5)," *TV-Sat-Magazyn*, September, 1989: 7.
[88] Ibid.

I have included an extensive excerpt from Sylweriusz Ładyński's article because it contains a very specific neighbourhood satellite TV schedule, and therefore, tells us much about the tastes of Łódź's inhabitants in the late 80s. Equally interesting, and certainly more significant, is the suggestion arising from this text that the viewers considered satellite television as an area of unconstrained freedom. Therefore, any change in schedule was seen as a form of censorship, typical of state television. It did not matter then, whether it was moral or political censorship that they were being subjected to. Either way, they felt that access to information was important. As such, they purchased satellite TV hoping for access to all available channels. At the same time, it is important to note that awareness of this liberating potential of sky television forced the members of the organising committee to self-limit the power that they, as operators and supervisors of the installation, exclusively held. Putting the matter briefly aside, a seemingly trivial fact of distributing satellite television through the AZART network set off fundamental questions about the democratisation of social life, and this in turn influenced the changes in the general political climate within the declining PPR.

Fig. 8. Andrzej Palma next to the antenna that was installed in the Manhattan estate in 1988

Source: Krzysztof Jajko, 2015

The success of Sylweriusz Ładyński's project resulted in two further initiatives, both aiming to build a network of collective satellite TV reception on Chojny Zatorze and Śródmieście/Manhattan estates. Since most of the actions pertaining to the construction of a satellite network in those two areas did not differ substantially from the actions of the "Main Organising Committee" on the "Piastów" estate, I shall focus my description on the exceptional elements of each aforementioned venture.

Neighbourhood television on the Chojny Zatorze estate can be described as Sylweriusz Ładyński's illegitimate child. This is demonstrated by the following statement from Paweł Bruś, who supervised its construction:

> At that time of [19]88, we were delighted to find out that on a neighbouring ["Piastów"] estate two completely uncensored satellite channels were being received. In response to this news, we organised a residents meeting, where we let everyone know about this, in order to gain our fellow residents' approval for a similar venture on our estate.[89]

It is important to note too, that from the very beginning of the work on launching a second neighbourhood television network in Łódź, Bruś remained in close contact with Ładyński, seeking his advice on both technical and organisational issues.

Moreover, Bruś could count on professional help from several residents of the Chojny Zatorze estate, who eagerly supported his initiative after the first residents meeting:

> First of all, we put together a team of 3–4 people to get started with the work of organising everything. In turn, we took on the technical side to find out how the signal, received by one satellite antenna, could be distributed to individual buildings. As it turned out, amplification was the answer.[90]

It follows from Bruś's account that the members of the team were able to complete this task thanks to their previous work experience:

> I was working at Telekomunikacja Polska [Polish Telecommunications] at the time. My colleague Bednarski was employed as an electrician in textile manufacturing. In a way, we had the technical background for the work that needed to be done.[91]

---

[89] Paweł Bruś, interview, November 27, 2015.
[90] Ibid.
[91] Ibid.

The Porion company also played a big part in the venture, importing satellite TV reception equipment, broadband amplifiers, and Ericson concentric cable from abroad. Its authorised representative installed the antenna on the roof of one of the buildings and set up a broadcast studio in the same building.

From the very beginning, the members of the local organising committee were realists confident that with persistence, they could accomplish what they wanted to:

> Our housing estate is large, counting 105 buildings. That is 105 blocks of flats, distributed across a large geographical area [...]. At the very beginning, our idea was limited to, say, 3–4 buildings. Later [...] after we completed that first stage, set up the studio and distributed the signal over cable, the whole thing just exploded. People could see the quality of the broadcast and also that they'd be able to watch uncensored news programmes.[92]

The policy of small steps worked not only regarding the residents, but also the housing cooperative's officials, without whose support the entire operation would be futile. For a time, however, they approached Bruś's actions with reserve:

> In the beginning there was some resistance. But after the first building, and after satellite TV was installed in three neighbouring buildings, those people saw that we weren't doing anything wrong, we're not devastating [the buildings].[93]

A sign of support from the cooperative's officials for the bottom-up initiative of its members was the fact that after some time the chairman of Chojny Housing Cooperative's Supervisory Board took a place in the neighbourhood television's organising committee.

However, the cooperative's participation in the construction of a collective satellite television reception network was not limited to symbolic gestures only. It offered tangible assistance, especially with raising funds, when it provided the committee with a bank sub-account. Taking on the financial management of the entire venture, the cooperative not only unburdened the organisers of a considerable part of organisational duties, but also through the power of its authority, gave more credibility and transparency to all transactions and accounts. In Bruś's opinion, this turned out to be incredibly important when upon the installation's completion, its creators' particular actions came under review:

---

[92] Ibid.
[93] Ibid.

[...] I suppose that the police took an interest in this later. I know, because I felt it. [...] After all, there were people living on our estate who worked for the police, and we'd see them. They'd be like, "you've bought dollars, what happened to them?" But I was doing this so transparently, that our finances could be checked at any time. Besides, by transferring funds to the cooperative, we were showing that nothing bad was being paid for with our money.[94]

The above statement is a perfect illustration of the complexity of the situation for the satellite television network creators who were working on the Chojny Zatorze estate. Aside from organisational and installation works, they had to take part in a complicated game with the communist authorities, specifically with employees of the Ministry of Interior. On one hand, according to other statements by Bruś, the police did not interfere with the network's development, even in spite of the fact that it had legal grounds to do so (as the venture's initiator had, in the beginning, acquired a permit only for individual satellite TV reception). On the other hand, the Ministry of Interior's representatives were officially obligated to control any bottom-up social initiatives, and to find illegality within them, such as uncovering black market currency transactions, which were forbidden in the PPR.

This dichotomous attitude of Łódź's police was undoubtedly a reflection of the communist authorities' twisted policy; in search of a lesser evil, the collective satellite TV reception was assumed to be a better solution than individual access, which was impossible to control en masse. Nevertheless, the authorities' leniency towards Łódź residents' thirst for new media can be explained in a different way as well, as Bruś points out:

Łódź was considered a red city. But it wasn't the case, that it was a red city. It was a worker city, but not a red one. [...] In my opinion, the authorities had to allow for some kind of outlet at that time, because if it could not provide meat, could not provide sufficient amounts of cold cuts, then the least it could do was allow people to watch some of the things people have around the world.[95]

It is difficult to say if the Party was managing such a measured and calculated image campaign in the late 80s. The final effect of the authorities' permission for neighbourhood television development was that citizens of a parochial socialist country could, for little money, experience the pleasure of consuming the audiovisual output of the West:

---

94 Ibid.
95 Ibid.

The first impression was to gasp at the visual quality. This was something complete-ly different. There were no reflections. There were no pauses in transmission. [...] Nature programmes were very interesting. We didn't have such brimming nature documentaries on Polish television. Eurosport was a must-have, because at that time Eurosport was transmitting everything that was happening in the West but wasn't available to us.[96]

Nonetheless, there are two sides to every story. As already mentioned in the description of Ursynów's experiment, the media gratification of satel-lite TV access did not push Poles into a state of civic inactivity that is typical for consumption of mindless television. Instead, it activated them, diminishing the power of social atomisation, which had been convenient for the government. The construction of local television networks was usually accompanied by agita-tion, awakening the will of cooperation and mutual help in the so-far apathetic citizens of the PPR:

We could only do this work once our normal working day was finished, so we'd usually go up at four o'clock in the afternoon, and sometimes we wouldn't get down from the buildings until four in the morning. If we were going up on a building at two o'clock at night, then nobody was asleep in that building. It seldom happens that people are so engaged socially, but these people wanted to help, and so, we never felt at an inconvenience doing this work.[97]

It is worth noting that this civic mutual help would sometimes take an inter-esting turn. According to Bruś's account, a well-known journalist from Łódź used his press connections to acquire a TV signal meter for the installers, unavailable in the city at that time.

Andrzej Palma, creator of the satellite TV collective reception installation on the Manhattan estate, points out the civic dimension of neighbourhood tele-vision as well:

In mid-December [of 1988] we were practically ready for the installation. The pres-sure was on to finish in time for Christmas Eve. As a result, the number of phone calls [was huge]. [People were saying]: what help do you need, what needs to be done, I'll get it, I'll help. There was something wonderful about it. I remember that at that time, after the success of "Solidarity," everyone was waiting for something to start happening in Poland, and yet nothing was happening. All the time, we were just waiting for those new events. That motivated everyone to get stuck in.[98]

---

[96] Ibid.
[97] Ibid.
[98] Andrzej Palma, interview, October 28, 2015.

This potential for social mobilisation among Poles was already evident in their coming together to make neighbourhood television happen. As Elżbieta Biwan-Kwiecińska – a member of the Śródmieście Housing Cooperative's supervisory board – reminisces, many people engaged in the creation of the enterprise, offering different kinds of help:

> A group of engineers came together under Dr Korczyński's leadership, who at that time was an assistant professor at our Technical University. There was Mr Andrzej Palma [as well]. They were knowledgeable about this kind of thing so we relied on them as the professionals. One our other residents, a lawyer called Brodniewicz, also offered assistance. None of them charged for their services, because all of this was done for the public good. They were real social activists.[99]

Everyone got involved when the organisers began the works between 22 and 23 December:

> People were watching us then from their windows. Many [people] would visit us; all those initiators, those who were collecting [money]. […] Even problems like getting good drills for concrete – because it was reinforced concrete everywhere – were solved by the residents. There weren't many such tools back then, so someone would bring us their drill from work. The amount of people who got stuck in was extraordinary, everyone helping to get the system up and running before Christmas.[100]

Such a short construction time in building number 231 deserves recognition, because installing neighbourhood television on the Manhattan estate, where all the buildings are around 20 storeys high, posed much greater difficulties than in a usual block of flats. This pertained, especially, to the installation of an antenna receiving signal from space:

> We chose the biggest antenna in existence at that time – three metres in diameter, a full parabola. This posed a problem, because it wouldn't fit in any of the lifts since the doors were less than two metres wide. Given that this was going to be our biggest challenge, we enlisted the help of our friends from the Alpine Club of Łódź, who built a safeguarded climbing wall. As we hauled the antenna up to the 21st floor, two climbers made sure it didn't hit any of the windows, which was a real danger especially as it was windy. [101]

---

[99] Elżbieta Biwan-Kwiecińska, interview, October 28, 2015.
[100] Andrzej Palma, interview, October 28, 2015.
[101] Ibid.

No problem they faced was greater than the issue of distributing the concentric cable inside the buildings, however. Like Biwan says:

> The height in certain places there is just half a metre and there are already wires, cables, hot water pipes and such. It was terrible physical labour, because you had to crawl on your stomach, especially in the attic, to pull those cables through. They did it very quickly though, very efficiently.[102]

The enthusiastic reception that neighbourhood television was met with among the residents of building number 231 (and elsewhere), assured its creators that their efforts were not in vain:

> It was a success that we got it up and running and for the residents, it was the most wonderful Christmas. People even came from neighbouring buildings to see what this television looked like and what kind of channels and programmes were showing.[103]

It is worth pointing out that the most important TV channels of the 80s were available in Manhattan from the very beginning:

> People were going mad for the 24-hour news channel CNN. I remember Canal+ [still under the name "FilmNet" at the time] was also popular. These two channels reigned supreme. Many people, however, submitted [suggestions], to broadcast [a channel] in German for example, even for some part of the day, because a lot of people knew that language, and later a French one as well. It was as if everyone wanted to test their foreign language skills.[104]

To a large extent, these requests to expand the channels offered stemmed from Manhattan estate's specific social structure:

> For so long, there was only this official state television, and then suddenly, these foreign channels appeared. For the residents, this wasn't a problem however, as many of them were intellectuals, who could speak these foreign languages.[105]

People's enthusiasm for satellite television on the Manhattan estate peaked on New Year's Eve 1988:

---

102 Elżbieta Biwan-Kwiecińska, interview, October 28, 2015.
103 Andrzej Palma, interview, October 28, 2015.
104 Ibid.
105 Elżbieta Biwan-Kwiecińska, interview, October 28, 2015.

It was probably the coolest New Year's Eve party in Łódź. [...] People were inviting their friends from across the city to show them what satellite television was all about. It was marvelous fun, because people were out on their balconies, setting off fireworks. There's [...] a beautiful green plaza here, and after drinking champagne the entire company came here and celebrated together. [...] And that perhaps, was the driving force for Łódź, showing people that things are possible.[106]

Considering that just six months later Polish society gave the communist government a red card in the parliamentary elections, this lively New Year's Eve held under an exterritorial firmament has to be considered one of the first stages of pulling down the Iron Curtain and opening Poland up to the West. This was expressed best by a phrase recalled by Biwian, which sums up the residents' reaction to the news that satellite television would be made available on the Manhattan estate: "Wow. Now that's a whiff of Europe."

However, even without taking into account this broader context of political changes that took place in Poland after 1989, it has to be noted that the satellite initiative on the Manhattan estate, similarly to Ursynów's experiment and projects on "Piastów" and Chojny Zatorze estates, were evidence of a resurgent community spirit among Poles. Examining the matter on a micro scale, it can be argued that because of neighbourhood television in the 80s, cooperative self-governance started working again in Poland.

The decline of involvement in local community matters among residents of urban estates in the PPR arose from many factors. First of all, following the influx of vast numbers of farmers torn out of their natural environment into the cities after the Second World War, a phenomenon of "amoral urban society" quickly set in, characterised by an absence of social ties in newly created urban areas.[107] Secondly, the city residents saw the estate's administration as a part of the communist apparatus and assumed that any involvement in the cooperative's life would not have any influence on its decisions in the end.[108] Thirdly, the creation of organised communities was made difficult by the big size of city housing estates built since the 60s.[109]

The processes presented above were largely directed by the communist authorities, who

---

[106] Andrzej Palma, interview, October 28, 2015.

[107] Jacek Tarkowski, *Władza i społeczeństwo w systemie autorytarnym* (Warszawa: Instytut Studiów Politycznych PAN, 1994), 276.

[108] Arkadiusz Peisert, *Spółdzielnie mieszkaniowe: pomiędzy wspólnotą obywatelską a alienacją* (Warszawa: Wydawnictwa Uniwersytetu Warszawskiego, 2009), 114–115.

[109] Ibid., 113.

saw a danger in independent cooperatives, as they were an important link in the self-governance of society (civic society, we'd say today). Because of ideological kinship, however, they could not or did not want to combat them openly. The authorities have thus striven to dominate the cooperatives on the one hand, but preserve their cooperative facade on the other, in order to engage society's energy and means.[110]

In the case of housing cooperatives, various community action initiatives had such a facade. Like other initiatives of this kind, they were unpopular among Poles. As sociological research from the 70s shows, only schoolchildren took part in neighborhood initiatives, forced to do so by their teachers.[111]

Things were different when it came to local television organising committees. Meetings convened by them attracted thousands of people, and the management and installation works that followed were met with great enthusiasm from residents of particular buildings. They committed to raising funds and offered to help. Therefore, in contrast to the institutional social ties that the authorities were artificially (and ineffectively) trying to create among Poles through community action, spontaneous involvement in creating neighbourhood television from scratch, forged real ties between people. At this point it is once again worth recalling an extract from Marek Kasz's editorial reporting on the social involvement of Warsaw's residents in the Ursynów experiment:

> Even though it is not here yet, satellite television has already brought about positive change. First and foremost, it has compelled many thus-far apathetic residents to band together and start collecting signatures. These are people who, like me, could not be convinced to join community projects to convert waste land into parks and tennis courts. Now, they are quite happy to be harassing their caretakers to compile lists of residents wanting to install satellite television, and to make them do it on time.[112]

* * *

The history of constructing three neighbourhood television networks in Łódź proves that Franciszek Skwierawski's critical remarks aimed at such social initiatives were over the top, particularly since stories about Łódź's satellite TV collectives continued to circulate, in some cases up until the 2000s. Paweł Bruś and his co-workers opened a private company right after finishing the installation on the Chojny Zatorze estate in 1990, building local television networks in other parts of Łódź as well as in other cities. In turn, by 1993

---

[110] Ibid., 100.
[111] Ibid., 116.
[112] Kasz, "Satysfakcja."

Andrzej Palma had connected to his network most of the buildings around Piotrkowska and Piłsudskiego streets, and opened the Cosmos TV company, which provided satellite signal to the residents of the Śródmieście estate, up until 2006.

A window to the world for Poland B[113]

All the examples of satellite TV dissemination described in this chapter so far took place in big cities. This does not mean, however, that the phenomenon of satellite television in Poland only went as far as the outskirts of Warsaw or Łódź. Since the cosmic signal was covering the entire territory of Poland, anyone, even a resident of a small village, could enjoy the benefits of television from the sky. The barrier was therefore not geographical, but monetary. As discussed earlier, access to receiving equipment and antennas was severely limited during the initial stage of satellite television's development in Poland, mostly due to the small number of salesmen importing merchandise from the West. Because of this, prices of satellite TV sets remained high for a long time and were converted into foreign currency. As such, only the affluent could allow themselves access to satellite channels, whether they were living in big cities, smaller towns or even villages in the outermost regions of the country.

The accuracy of this claim is proved by an article titled "Komu program z satelity?" [Who wants programmes from satellite television?] from 1988, published in a *Nowiny* daily from the Podkarpacie region

> Not long ago, satellite television was something very distant from us. Today, it's knocking on our door louder and louder, and not only through the Second Channel's programme *Bliżej świata* (*Closer to the World*). In the National Radio Inspection's regional office in Rzeszów […] 11 individual satellite TV reception permits have been issued. Eighty people were given permits for purchase of satellite reception equipment, and around 1,200 applications are awaiting examination.
>
> Among those who have satellite dishes on their rooftops, which allow for receiving international content, there are, for example, three farmers, a teacher, a priest, and a private businessman. There are cultural and education facilities as well. These chosen ones already have the possibility to watch, among others, the Swiss TELECLUB and Italian RAI UNO.[114]

---

[113] Poland B is a term describing a less developed part of the country situated east of the Vistula River.

[114] Janusz Pawlak, "Komu program z satelity," *Nowiny*, September 27, 1988: 6.

Admittedly, the group of individual satellite TV users who lived in the Podkarpacie region in the late 80s was not very impressive in number and nor was there much social class diversity. But this was the case everywhere. In Łódź and Warsaw, very few owned private satellite TV reception equipment. Considerably more interesting is the fact that cultural and education facilities had access to cosmic signals at that same time. As other press reports and witness testimonies illustrate, the phenomenon of satellite TV dissemination through such institutions was very common in the Podkarpacie district.

Probably the most interesting case to examine is the launch of the collective satellite TV reception installation in the Regional Cultural Centre in Tarnobrzeg. Not only did it involve a pioneering technical innovation, but was also accompanied by remarkably interesting promotional activity, leading with an event that, in May 1988, was spotlighted even in the nationwide magazine *Ekran*:

> The satellite television family was recently joined by the Cultural Centre in Tarnobrzeg. The first screening took place there on 22 May, when a dozen foreign channels were broadcast across several screens. "Satellite Sunday," supplemented by video screenings, attracted hundreds of residents who had the chance not only to see the new means of communication, but also to listen to a speech [delivered by Franciszek Skwierawski] titled "Satellite television and VCR – present state, perspectives."[115]

A more detailed account of the event can be found in an article published in the local magazine *Tygodnik Nadwiślański*:

> A permanent satellite dish – the first in our region – is owned by the Sulphur Basin Cultural Centre in Tarnobrzeg. It allows for reception of 20 channels broadcast by three telecommunication satellites. Last Sunday, during an all-day event called "Satellite Sunday," two direct satellite broadcasts and three retransmissions of previously recorded programmes were shown on seven screens in the Regional Cultural Centre. A market was organised as well, where VCR and computer equipment, video recordings, cassettes and computer programmes were available for purchase and exchange.[116]

An important clarification must be made here. Despite what was written in *Ekran*, the first public screening of sky television in Tarnobrzeg did not take place during "Satellite Sunday." According to a recollection from Jarosław Piątkowski, the initiator of satellite TV installation in the local Cultural Centre,

---

[115] "TV-SAT w Tarnobrzegu," *Ekran*, June 23, 1988: 3.
[116] No title, *Tygodnik Nadwiślański*, May 27, 1988: 2.

the first signals from space were received in the early spring of 1988 (probably in March) and made available to everyone frequenting the institution shortly after.[117] It was made possible by an already-functioning, clever installation designed by Aleksader Dyl, another Regional Cultural Centre employee, who was operating an amateur video club there. Here's how its construction came about and what purpose it initially served:

> I came up with the idea that I'll fashion a kind of television network here in this Cultural Centre. The television network consisted of several monitors placed around the café, in the hall and in my workshop [...]. There were monitors and CCTV [included in the installation]. For example, there were cameras showing the stage in the Cultural Centre. [...] The idea was to attract the onlooker, to get him interested in what was happening on the stage. [...] [Moreover] we were making announcements using the CCTV and a very primitive VCR (one using reels, not cassettes). We were making a sort of television studio. [...] In breaks or before a show, content was played on these VCRs, with an announcement of the upcoming show (what can be expected of it). In the meantime, during the intermission, interviews with actors would be screened. [...] And from this artistic activity we moved on to a more commercial one. So we were simply serving this Cultural Centre.[118]

When an almost two-metre dish was installed on the Cultural Centre's roof in early 1988 and began receiving the first signals from the Eutelsat satellite, the installation described above (after many modifications) was assigned to distribution of foreign TV programmes. Here's what Jarosław Piątkowski says on the topic:

> We didn't want to shut ourselves in with this decoder, with this television in a small room in the Cultural Centre (because all of that was in such a small room, literally maybe 4 square metres). We wanted for people to see it, so we were looking for ideas on how to go about that. There were no projectors at the time, so we started organising [special screenings]. [We had] a large television – it was 25, maybe even more, 27 or 29 inches, and we would place this TV in the main room. You could walk around and watch at the same time. There was MTV, then SAT 1, then something else. There was always something on. Later on, we installed TVs throughout the Cultural Centre and you could watch specific programmes on many televisions in different parts of the building. In the beginning, these TVs in the corridors were black and white. Later, colour TVs were brought in. That was fantastic.

---

[117] Jarosław Piątkowski, interview, March 2, 2015.
[118] Aleksander Dyl, interview, February 28, 2015.

[…] we'd created this system across several televisions in several rooms. You could watch these broadcasts every day, say, in the evening, drinking coffee and watching MTV, or some German channel. And back then, obviously, no one had any way to have things like that so that's how we were showing this bit of the world to people, to residents of Tarnobrzeg.[119]

Fig. 9. Tarnobrzeg citizens get acquainted with the satellite television

Source: Wacław Pintal

[119] Jarosław Piątkowski, interview, March 2, 2015.

According to Piątkowski's account, the most popular channels among Tarnobrzeg's youth especially were the music channels:

[…] Szewczyk and Pijanowski were showing up to five music videos during their
TV programme, broadcast once a week.[120] Now there was MTV, broadcasting
24 hours a day, which we could watch whenever we wanted. It was an unbelievable
experience.
       […] The biggest crowds were made up of young people, who knew that be-
tween four and eight o'clock those TVs were going to be showing music channels.
It was a voyeurism of sorts, because you didn't have that at home every day.[121]

Because Piątkowski was a music man and the "musical window to the world"
created by satellite television was especially valuable in his eyes, he started
the trend of watching music videos in the Cultural Centre:

There was one additional thing we were doing. In the Cultural Centre there was
a TV in the window. We'd turn the volume up high so the sound of cool, foreign
music could be heard by people passing by. Whatever the weather, they could walk
in and take a look. There were three or four such TVs and people could simply pull
up a chair, sit down and watch for a bit.[122]

Interestingly, music videos were not only watched live, but also recorded
with the use of video equipment available in the Cultural Centre. Compilations
created this way on VHS cassettes had a regular, but at the same time peculiar,
circle of recipients:

There were people working with the Cultural Centre: DJs who, in addition to their own
sets, introduced a television element. This was a bit weird though. A DJ would stand
there playing his records with two TVs next to him playing something completely different. Later on, there were even auto-plays of a kind, because MTV had these night
[programmes] playing a mix of the biggest hits. Since we had access to that, even this
little world of music and DJs around wanted to see what's currently on top.
       […] This didn't suddenly turn into an industry although there were music peo-
ple and some music journalists who came by from time to time […]. We weren't
limiting this for anyone too as the law did not forbid us from copying this type
of content and distributing it further. After all, that's why we were there, to help
people gain access to music culture, at least, to some degree.[123]

---

[120] Krzysztof Szewczyk and Wojciech Pijanowski hosted the television show
Jarmark (*The Fair*) in the 1980s.
[121] Jarosław Piątkowski, interview, March 2, 2015.
[122] Ibid.
[123] Ibid.

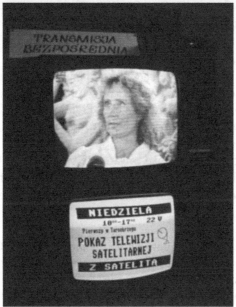

Fig. 10. Satellite installation in the Cultural Centre in Tarnobrzeg

Source: Wacław Pintal

Other popular video recordings of satellite TV programmes had completely different viewers, who, today, we'd call members of the middle class:

There was this programme, it doesn't exist anymore, called *Landscape*. This was a kind of showcase of the most beautiful places from around the world, with classical music playing in the background. A sort of weird music video. […] There was no commentary, no information. Sometimes we would record such things, because more and more people had VCRs. They were even asking us for it: "Record something like that from the satellite for us."[124]

These local practices of watching VCR recordings of satellite TV programmes were never developed on a mass scale. Although the VCRs presence in the Cultural Centre made it possible to record and distribute Western movies absent from Polish cinemas, Piątkowski and his coworkers decided not to make such a move. This decision came mostly from the fact that the movies provided through satellite were broadcast in the original language. This fact, combined with a low level of knowledge of foreign languages in the PPR, would necessitate supplementing the copies with a Polish translation, an activity that went far beyond the work duties of the Centre's staff. As a result, during the following months (and even years) the Tarnobrzeg residents watched satellite television live on the premises of the Cultural Centre. This is evident from the following press release published in the *Tygodnik Nadwiślański* in late 1988:

> Satellite TV screenings organised by the Cultural Centre in Tarnobrzeg are gaining in popularity. Across five screens situated in places most accessible to the audience, around a thousand people watch a programme every day, selected on the basis of information published by *TOP* weekly. Eighteen channels are currently being received, broadcast every day between 9 am and 9 pm.[125]

Tarnobrzeg's positive response to satellite television at the Cultural Centre meant that it was a huge success for the young organisers of the whole venture. Even so, acquiring permission and funds for installing the satellite TV receiver set was no easy task. As Piątkowski reminisces:

> We were trying to, quote-unquote, pester the director [of the Cultual Centre] about this. For a while all she would say was, "No, no, no, no. " Later on, the funds were found somehow, meaning someone must have decided to finance this, because there were no grants or subsidies at the time. I don't know who made this decision that money must be given to pay for Poland B, if not C's access to the wider world, but it really was a grand event for the region.[126]

It follows from other statements by Piątkowski that the main "sponsor" of the satellite initiative in the Regional Cultural Centre in Tarnobrzeg was the Voivodship Office's Culture Department. Aside from that, Siarkopol, a powerful

---

[124] Ibid.
[125] No title, *Tygodnik Nadwiślański*, October 21, 1988: 2.
[126] Jarosław Piątkowski, interview, March 2, 2015.

nationwide company extracting and processing sulphur operating in the Tarnobrzeg area, also offered the young culture and education workers large sums
of money. It was not a coincidence that the full name of Tarnobrzeg's culture
and education institution was Sulphur Basin Regional Cultural Centre.

Fig. 11. The initiators of satellite television in the Cultural Centre in Tarnobrzeg

Source: Wacław Pintal

In the years that followed, the Culture Department and Siarkopol's involvement in the dissemination of satellite television in the region went beyond the occasional patronage of the Cultural Centre's spontaneous initiative. According to
a short press release published in *TV-Sat-Magazyn* in 1989, Siarkopol was planning to launch professional cable television in the city and the surrounding area:

The Tarnobrzeg region will have satellite and cable television. On the initiative
of sulphur mine and processing plant "Siarkopol" in Machów, foreign currency has
been consigned for this purpose and contracts with reputable Polish and foreign

companies have been drawn, with the first batch of equipment already purchased. At first, satellite television will be broadcast in the city's cultural institutions and sanatoriums. After that, cable television in residential areas will be installed.[127]

Although this remarkably ambitious project never came to be in the end, it does not change the fact that in the late 80s Siarkopol was the only institution that could take on the organisation of cable television in Tarnobrzeg. This was mostly about possessing the foreign currency mentioned in the press release. After all, in the 80s, Siarkopol was one of the country's biggest exporters.[128]

However, a different initiative, put forward at the same time by the Voivodship Office's Culture Department in Tarnobrzeg, took shape. Here's what Piątkowski says on the subject:

For two years it was crazy. Those two years are also what led to the whole region learning about satellite television. Seeing the interest in Tarnobrzeg, the Voivodship Office's Culture Department decided that it would like to equip other cultural centres in the former Tarnobrzeskie district. Maybe not as big, maybe a bit smaller, but still providing that window to the world. I became a sort of coordinator for this project. Two years later the Voivodship Office funded [satellite television] in several cultural centres, eight in total, I think.[129]

Piątkowski himself tried to investigate the reasons for such lenient cultural policy towards ideologically alien audiovisual attractions from the West:

Apparently, they did not see this as an enemy of the system, since they wanted [this] installed in other centres as well. Quite the opposite, they probably wanted to show how cool they were.[130]

This generous gesture from the Culture Department (but also from Siarkopol) can be interpreted in a different way as well, especially when taking as a point of reference the initiatives connected with the propagation of satellite television on a national level. This alternate interpretation focuses on the activity of the so-called satellite television clubs. Similar to "cable televisions for the poor," these associations, functioning mostly at or around workplaces, were to constitute the second method of satellite television's ideological familiarisation through

---

[127] "Satelitarna i kablowa TV dla regionu tarnobrzeskiego," *TV-Sat-Magazyn*, August, 1989: 4.

[128] Tomasz Sandomierski, "Parada kolosów," *Polityka*, supplement: *Polityka – Eksport – Import*, February, 1985: 1.

[129] Jarosław Piątkowski, interview, March 2, 2015.

[130] Ibid.

collective reception subject to external control. Proof of the communist authorities' involvement in the execution of the above strategy is the fact that in November 1988 there were already 15 such clubs functioning in Poland, under the auspices of the *Audio-Video* magazine.[131]

The fact that the satellite TV installations funded by the Culture Department in the Regional Cultural Centre in Tarnobrzeg and other cultural centres in the regions fitted, to a certain degree, to the satellite television clubs' scheme of operation is evidenced by Piątkowski's statement describing how, during the Polish Round Table Talks, the police had sealed the equipment, preventing reception of the American news channel CNN. Significantly, when recalling this event during his interview, Piątkowski uses the expression "martial law" instead of "Round Table" in a slip of the tongue. Admittedly, he corrects his mistake later in the conversation; however, it might be the case that he subconsciously expressed his emotional attitude towards the entire situation in this way. In short, blocking satellite television by the police was, for him, an action tantamount to the censorship used by the communist authorities between 1981 and 1983.

There's one more potential explanation for the policy of the Culture Department. It can be inferred that local authorities were trying to follow the new trends in socialist cultural policy apparent since the beginning of the 80s when they provided residents of the city and its vicinity with satellite television through cultural centres. A general outline of this refreshed approach to culture promotion can be found in a book by Władysław Misiak, who, on the basis of empirical research pertaining to the social practices of Lublin's inhabitants, puts forward specific conclusions for the future:

> While the decisions and processes of municipal bodies remain formalised and mostly unchanged, the cultural life of the city is evolving spontaneously, which P. Rybicki writes is a reflection of changing cultural values. As such, basing decisions about the city's future on old value systems does not produce the right results.[132]

Moreover, Misiak states that:

> [...] limiting municipal bodies' involvement in cities' cultural life is encouraged. Research conducted has proven that excess formalisation of the system and institualisation of culture curtail city development, as does a simplification of cultural

---

[131] DJB, "Kluby TV-SAT," *Audio-Video*, no. 1 (1989): 24. It is worth noting that the activity of the satellite television clubs was described in the *Audio-Video* magazine in a style resembling the press commentaries on the Polish cultural revolution in the fifties: "All the clubs decided to install the equipment they had on their own. The stronger clubs promised to help the weaker ones."

[132] Władysław Misiak, *System kulturowy miasta uprzemysłowionego* (Wrocław: Zakład Narodowy im. Ossolińskich – Wydawnictwo, 1982), 206–207.

goals for the future. In light of these conclusions, it is imperative that cultural workers and educators possess adequate knowledge and skills, especially in industralised cities. As the study establishes, the economic and social situation, as well as cities' demographics and people's moral values are changing every few years. As a result, the skills educators and cultural workers must possess are ever-shifting and thus, their ongoing education is necessary[133]

Because of that, the author states explicitly elsewhere that:

Our recommendation is the early detection of newly appearing trends and "qualitative mutations" of a cultural subsystem, usually initiated by a small group of city residents. Careful observation of residents' diverse needs is necessary, as is paying attention to their behaviours. This must be followed by theoretical reflection.[134]

It is difficult to say if the officials in the Culture Department in Tarnobrzeg had an inclination for "theoretical reflection," but they could certainly react to what was in front of them. Thousands of people who visited the Cultural Centre every week to watch satellite television were a model example of a "newly appearing trend," that the decision makers caught just in time. As a result, a decision was made to install additional satellite TV reception sets in several cultural centres in the region.

It is important to note, however, that none of the things described above would have materialised if it were not for the proactive and ingenious work of the initiator of the entire venture, Jarosław Piątkowski, and his coworkers Aleksander Dyl and Wacław Pintal. It was them, probably due to their young age, who sensed new trends in how media could be used for the popularisation of culture, and thus they were the first culture workers in Tarnobrzeg to respond to broader lifestyle changes taking place in late 80s Polish society.

A similar suggestion had already been made by Józef Myjak in 1988, in the *Tygodnik Nadwiślański*, as a summary of local culture:

The installation of a satellite dish on the roof of the Sulphur Basin Regional Cultural Centre in Tarnobrzeg was an indisputably revolutionary event. This great invention of the recent decades introduces a lot of changes to interpersonal communication, altering it radically and making the world a "global electronic village" – as put by a Western theoretician of culture. In one day, Tarnobrzeg became a part of the global mass culture.

Today it is still difficult to predict what fruit the satellite television will bear, especially for culture in the narrow sense. Won't the river, or even flood, of information

---

[133] Ibid., 202–203.
[134] Ibid., 204.

bring us art in a pill? Are the regions going to preserve their identity? Won't it destroy people's willingness to read, and consequently, thought and artistic contemplation, for which there simply won't be enough time? After all, tens of simultaneously broadcast channels are tempting and bewildering, as some mysterious, magical force makes us look into an electronic box. As usual, humans have the ability to choose what they want to watch, but maybe we should base our choices on what those with adequate knowledge recommend; those working in education or the arts.[135]

Let us add only in conclusion that in the same year another local magazine, *Nowiny*, was conducting a project across the entire Podkarpacie area called "Youth News Club." It consisted of one-day events organised in cultural and community centres, and the main part of their programme was a satellite TV screening.[136] In Poland B, more windows to the world were being opened ...

A guide to the cosmos of television

Undoubtedly, the biggest obstacle on the way to the joys of satellite television for many Poles was the narrow network of distribution. This does not mean, however, that someone who bought an antenna and a receiver could enjoy satellite television straight away.

First, the antenna had to be installed and calibrated to the right satellite. Naturally, such services were provided by representatives of companies selling satellite TV equipment, but that automatically increased the cost of the entire investment. For example, in 1987 the Baltona company was offering a Polish satellite dish for 150 thousand złoty, while its installation cost an additional 40 thousand.[137] However, considering that the cost of an entire satellite TV reception set could reach up to 8 million złoty, the price of installing an antenna was comparatively small. The problem lay elsewhere: namely, most of the antennas available on the Polish market had fixed mounts which, unlike polar mount antennas, did not allow for controlling the dish remotely in order to point it at different broadcasting satellites. When additional satellites started appearing in the sky every month, satellite TV set owners faced a serious dilemma. Either they were to acquire the knowledge on calibrating satellite dishes themselves, or they would have to call a professional every time they wanted to watch something from a different transmitter, paying a hefty sum for the service. The staff of magazine newsrooms,

---

[135] Józef Myjak, "Bilans w kulturze," *Tygodnik Nadwiślański*, January 6, 1989: 8.
[136] See series of press commentaries in the daily *Nowiny* from January to April 1988.
[137] tł., "Inwazja z kosmosu (1)," *TOP*, December 11, 1987: 6.

as well as authors of numerous guidebooks, quickly came to the aid of satellite enthusiasts troubled by such situations.

One of the first examples of a tutorial for enthusiasts of cosmic television was a series titled "Satellite TV. DIY," published in *Ekran* from May 1989 onward. Its author, Seweryn Jacek Kobyliński, considered the ability to set an appropriate elevation angle as a crucial skill for seekers of the satellite experience. This task posed no significant difficulty if the reader owned an antenna with a mount equipped with a protractor. If this was the case, they could begin to penetrate the firmament straight away, detecting further broadcasting satellites. Unfortunately, a lot of antennas sold in Poland lacked this facility and the author was aware of this. Therefore, in the third part of the series, he offered the following idea to the readers:

> If the antenna mount is not equipped with a protractor, then a device measuring the angle of elevation can be made on one's own. The device consists of a wooden slat with a school protractor attached with thumbtacks. A thumbtack is inserted in the middle of the protractor, with a small weight on a thread attached to it. After placing the slat to the upper and lower edge of the dish, the angle of elevation is read directly from the protractor's scale. The device is quite accurate, and its imperfection is that it can only be used during windless weather, since otherwise the thread with the weight dangles in the wind.[138]

Such publications presenting home-made solutions to satellite television problems during the times of socialist shortage could have given an incidental reader the false impression that audiovisual signal from space could be captured just as easily as terrestrial television. Given this misunderstanding, the author of the guidebook *Mój telewizor* (*My TV*), Konrad T. Widelski, was forced to get back to basics:

> A question arises: can a parabolic antenna for satellite reception be made on one's own, similarly to a dipole antenna for receiving typical "ground" TV transmission? The answer to this question is unfortunately no. The cap of the parabola must be formed with very high precision, which is beyond the skills of an amateur.[139]

The introduction of reception equipment of a higher standard in the early 90s, including remotely controlled antennas with polar mounts designed for receiving signal from several satellites, put an end to all the talk about satellite television for

---

[138] Seweryn J. Kobyliński, "Przygotowanie zestawu odbiorczego," *Ekran*, June 15, 1989: 30.

[139] Konrad T. Widelski, *Mój telewizor* (Warszawa: Instytut Wydawniczy Związków Zawodowych, 1989), 80.

the poor. In a book published in 1991 titled *Prawie wszystko o telewizji satelitarnej* (*Almost Everything on Satellite Television*), its author, Tadeusz Kurek, outlines the matter briefly: "The installation of polar mount antennas is very labour-consuming and requires experience. Therefore, it is worth leaving this task to the professionals."[140]

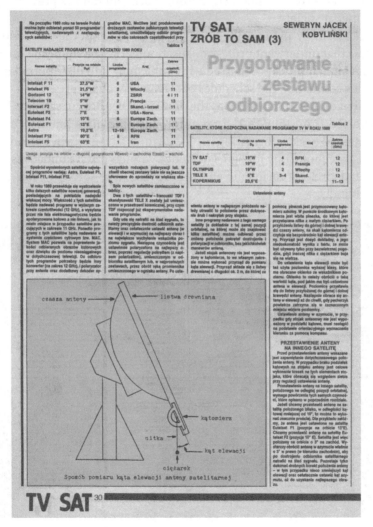

Fig. 12. One of the installments of the cycle "*TV SAT. Zrób to sam*" in *Ekran*

Source: *Ekran*, June 15, 1989: 30

---

[140] Tadeusz Kurek, *Prawie wszystko o telewizji satelitarnej* (Warszawa: Wydawnictwo Naukowe Techniczne, 1991), 25.

And so, the period of satellite experiments and amateur exploration ended with the advent of capitalism. Every year, more and more white dishes appeared on the balconies of houses and apartments, installed now by professionals employed across numerous satellite TV equipment stores. Ownership and operation of a satellite TV reception set irrevocably lost the quality of a romantic adventure, and recently thrilling images from space entered the grey everyday life of a consumerist society.

Yet the assembly and installation of antennas were not the only field of satellite business that underwent professionalisation in the advent of the Third Polish Republic. A similar process occurred with publishing and distribution of printed programmes containing the schedules of foreign TV channels.

Initially, activity of this sort was the domain of vendors of satellite TV equipment imported from the West. They printed simple fliers for their customers, which contained basic information on TV channels available in Poland. Enthusiasts would also photocopy foreign TV guides.[141] Such private initiatives, however, had a limited circle of recipients and most Poles were forced to watch satellite television blind.

Fig. 13. A cover of one of the many manuals for satellite television amateurs

Source: Tadeusz Kurek, *Prawie wszystko o telewizji satelitarnej* (*Almost Everything on Satellite Television*) (Warszawa: Wydawnictwa Naukowo-Techniczne, 1991)

---

[141] Henryk Ciski, "Między nami…," *TV-Sat-Magazyn*, May, 1989: 2.

It was most likely the staff of 1987 launched *TOP* magazine, who were first to recognise this gap in the market. From 23 September 1988, its pages contained a permanent column titled "TV from the Orbit," publishing a weekly satellite TV guide. It is important to note that the magazine's publishers were not compiling the column themselves, but rather used the services of the DIGITAL company, a satellite TV equipment distributor, who had access to the schedules of the most important foreign stations because of its commercial activity. In other words, thanks to being included in a nationwide, high-circulation magazine, satellite TV reception equipment salesmen's old method of clandestinely publishing timetables, spread on a mass scale. Such a cooperation benefitted both sides. The DIGITAL company, whose name was mentioned at the beginning of every schedule, gained free advertising, while the publishers of *TOP* magazine could count on an increase in sales.

The "TV from the Orbit" column's only weakness was the fact that very little space was allocated for it. It could not have been different, since *TOP* magazine was focusing on printing paid advertisements. This limited space was significantly affecting the column's shape, which the magazine's publishers were aware of. This is clear from the following excerpt from the issue published on 30 September 1988:

> Out of necessity (lack of space!) the recommendations printed in the *TOP* weekly are very condensed, containing several items in the most popular TV genres for the week. We have propositions of feature films, TV series, documentaries, children's programmes, news programmes, science and culture journalism, entertainment and sports. Hopefully they will be satisfactory, although we are aware that demand is growing.[142]

By the beginning of October 1988, a new method of presenting foreign schedules was adopted. It stepped away from dividing the timetable by channels, introducing genre categories instead:

> We're changing the way we publish the satellite TV guide. Facing the impossibility of printing it in full (lack of space), we've decided to recommend selected programmes grouped by genre. For example: feature films, TV shows, music programmes, sports, and so on, will be listed separately.[143]

In any case, these genological distinctions underwent numerous changes and simplifications in the following weeks. The aim was to fit as much content important to readers in as little space as possible. Finally, in early November that

---

[142] "Telewizja z orbity: Krótki informator TV satelitarnej," *TOP*, September 30, 1988: 29.

[143] "TV z orbity," *TOP*, October 7, 1988: 29.

year, the permanent genre layout of "TV from the Orbit" took shape. Its key categories, as they appeared in *TOP* magazine, were music and entertainment programmes, drama series, feature films, animated movies for children, documentary movies and series, sports, debate and news programmes. From a modern-day perspective, the genre arrangement of these programmes obviously presents excellent research material, through which the tastes of the first users of television from space can be reconstructed.

From time to time, the "TV from the Orbit" column went beyond the usual formula, focusing on one type of television programme such as a TV series:

> We start today's installment of "TV from the Orbit" in an unusual way. Because of numerous readers' requests, we're publishing listings for selected satellite TV channels. We'll begin with the SKY CHANNEL, because it is probably the most popular in Poland.[144]

At other times, one of the genres was omitted on purpose:

> In this week, we're, unusually, not providing a list of feature films. There are two reasons for this. First of all – these movies are almost regularly repeating every month. Secondly – simply switching to one of three channels: SKY MOVIES (ASTRA – between 5:00 pm and 1:00 am), FILMNET (ASTRA – all day) or PREMIERE (INTEL – 4:00 pm – 4:00 am) is enough to find an interesting feature film.[145]

As evidenced by the first statement quoted, these genre reshuffles were, to a large extent, dictated by readers' preferences. Moreover, they were demanding that not only the schedules fit their needs, but also wanted to read in-depth profiles of individual TV stations:

> Many of our readers, regular and occasional satellite TV viewers, are expressing exceptional interest in CNN, which is the flagship channel of Ted Turner's network. This may be because CNN programmes are widely popularised by Polish Television, and also because of the unique, 24-hour news channel's content itself. [...] Fulfilling the request, today we're publishing the so-called programming that is this this channel's typical daily schedule, separately for weekdays and weekends (Saturdays and Sundays).
> The presented schedule looks incredibly monotonous in print. It's a very misleading impression. Viewers of the "Dziennik Telewizyjny" ["Television Daily"] have a hard time imagining that 24 hours of news and debate can have their dynamics and drama. But they do. To find out for yourself, watch the CNN channel.

---

[144] "TV z orbity," *TOP*, February 10, 1989: 29.
[145] "TV z orbity," *TOP*, May 19, 1989: 29.

I am quoting such an extensive excerpt from the "TV from the Orbit" column from December 2 1988 because it illustrates the popularity of the American news channel, standing in complete opposition to reports from the domestic "Television Daily" on allegedly Polish attitudes. Moreover, the report's author allows himself to criticise the Polish news programme's formula. While he is criticising only the programme's format, as opposed to its content, such an open endorsement of a capitalist news channel is still quite astonishing to a contemporary reader.

Incidentally, it is worth noting that the editors of the "TV from the Orbit" column made many such opinionated digressions. Here is one of the most interesting comments, quite annalistic in nature:

> The "Closer to the World" programme broadcast on Sunday afternoons illustrates the benefits and pleasures of programmes broadcast by satellite providers such as Astra, Intelsat or Eutelsat. This certainly poses dangerous competition for our two domestic television channels, whose programmes are described by viewers as boring, although it must be added that there are many who hold the opposite view.
>
> In any case, the management of the Polish Television has to reckon with losing their monopoly with the presence of an incredibly attractive – although very costly – adversary. It is especially the case since 4 percent of Poland's geographical area cannot receive Polish Television's First Channel and 23 percent cannot receive the Second Channel. The reception quality isn't the best either, as opposed to satellite, which is excellent and available everywhere.[146]

Although the *TOP* magazine's journalists were trying to accommodate their readers' needs in the limited space of the "TV from the Orbit" column, they could not meet every demand and request made. The only way to do so was to create a special magazine, dedicated entirely to foreign TV programmes. It was created in Łódź in March 1989, under a title that left no doubt about its focus: *TV-Sat-Magazyn*.

The magazine's format was decided from the first issue. The opening pages, apart from offering an editor's commentary, contained interviews, letters from readers, minor reviews of selected programmes, as well as series consisting of several installments, like "The Satellite Vademecum." The rest of the magazine was filled with timetables of the most popular TV stations, interrupted every few pages with highlights from the world of fashion, film and music. Unlike in *TOP* magazine, a satellite television fan could finally find here a complete guide for broadcasters like TV 5, Sky Channel, MTV or RTL, enabling

---

[146] "TV z orbity," *TOP*, June 2, 1989: 29.

them to decide for themselves what would be most interesting to watch. It was undoubtedly a revolutionary solution that fundamentally changed satellite TV reception in Poland. It is worth adding, however, that the publishers chose to lure their readers in with cover images of scantily clad women and Playboy-like centrefolds, rather than risking running images of programmes that might only appeal to a niche audience. *TV-Sat-Magazyn* was an instant success, as illustrated by this excerpt from the editorial published in the second issue of the magazine: "The trial issue of *TV-Sat-Magazyn* disappeared from newsstands in cities like Poznań, Wrocław, Cracow, Katowice and Łódź in a matter of hours."[147] Moreover, the magazine's influence went beyond the country's borders, as explained in the June issue:

> As the vice-director of the Centre of Information and Polish Culture in Prague, Piotr Sagan, M. Sc., informed us, the *TV-Sat-Magazyn* was also to be found in Czechoslovakia, garnering considerable interest there. A few days ago, the executive producer of "Kamerat" (Vladimir Kuntz), a popular TV programme for young people, addressed the Counsellor of the Polish Embassy in Prague, Mirosław Rogulski, PhD, requesting the import of our publication for the benefit of Czech television.[148]

I do not intend here to analyse the contents of *TV-Sat-Magazyn* in detail, as I have in the case of *TOP* magazine. This is because it contained a great number of articles on satellite television and so, it would be difficult to give adequate space to the diversity of subjects discussed and multidimensionality of articles written by individual authors. I hope that the informative, as well as journalistic, quality of this magazine is attested to by other parts of this chapter, through quotes from such authors as Sylweriusz Ładyński, Andrzej Marciniak or the chief editor of *TV-Sat-Magazyn*, Henryk Ciski. However, I would like to devote a few words here to a peculiar phenomenon, namely crosswords included with the first issues of the magazine.

As usual in the case of such intellectual games, these crosswords were printed on the magazine's last pages. Interestingly, the crosswords in *TV-Sat-Magazyn* differed considerably from typical puzzles found in daily newspapers. Every one of them had to be solved in a different language (English, French, German or Italian), which was supposed to help readers in learning foreign vocabulary, useful for watching foreign programmes. So that no one missed this educational potential of the crosswords, their page was titled with the unequivocal headline: "We're learning foreign languages."

---

[147] "Drodzy czytelnicy!," *TV-Sat-Magazyn*, April, 1989: 2.
[148] "TV-Sat-Magazyn także w CSRS!," *TV-Sat-Magazyn*, June, 1989: 2.

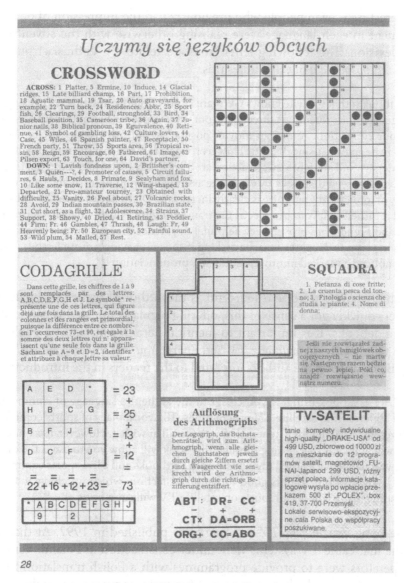

Fig. 14. A page with multilingual crosswords from the *TV-Sat-Magazyn*

Source: *TV-Sat-Magazyn* 2, 1989: 28

This is how we come to a third barrier, encountered by many early users of satellite TV: the fact that the programmes it provided were broadcast in foreign languages, and, unlike today, without a Polish voice-over. The fluency of Poles in communication codes of the West, which predominated in space television, was lacking in the period of the PPR.

Interestingly, even people who previously had the impression of being perfectly fluent in each language were realising otherwise with the advent of satellite television. The following letter from a satellite TV enthusiast published in the *Rzeczpospolita* daily, illustrates this:

> We thought that my wife knows German, I know English, and that we're using these languages quite fluently. A direct confrontation, however, has quite brutally corrected our conviction. American movies pose a big difficulty, same as German ones where people are using dialects. However, we've realised that the movies with lively, immersive and straightforward action are a great way of learning a foreign language. We've all agreed that week by week, the assimilability quotient, if you can put it this way, of foreign programmes, is clearly increasing. Dictionaries have been dusted off and returned to favour. An electronic translator with Polish, English, German, Italian and French, has proved very helpful. [...] A nice moment in this language education was our daughter's teacher asking if she was having private lessons, because she noticed a marked improvement in her pronunciation.[149]

Not all viewers were as resourceful as the author of the letter above. Book publishers quickly got a sense of this, offering regularly updated versions of language guidebooks. It is important to mention here, for instance, a publication under a telling title: *Zrozumiesz TV-Sat. Sport* [*You will understand satellite TV. Sport*]. Below is what the authors imparted to the readers in the introduction:

> *SPORT* is a book bringing you closer to the authentic language of commentators and athletes from satellite TV programmes. The vocabulary of sports commentary represents a barrier for listeners that aren't familiar with it. However, it is quite limited, which means that getting through this book will allow you to travel the distance between language learned previously from textbooks and a live sports commentary. [...] We wish our readers good fun and satisfaction, which watching and understanding many wonderful sports on satellite TV will soon bring you.[150]

The book *Zrozumiesz TV-Sat. Sport* was published in 1992. At that same time, the first cable TV systems were under construction in Poland. Very soon their operators were to provide programmes with a Polish translation to their subscribers. And so, the heroic period of the first conquerors of satellite television was coming to an end in the area of linguistic battles as well.

Translation: Wojciech Szymański

---

[149] Franciszek Skwierawski, "Ta bariera językowa..." *Rzeczpospolita*, October 12–13, 1991: 8.

[150] Bożena Pudlik, Stefan Świętochowski, *Zrozumiesz TV-Sat. Sport* (Gdańsk: Wydawnictwo Mariol, 1992), 5.

Fig. 15. An English guide for sport TV channel fans

Source: Bożena Pudlik, Stefan Świętochowski, *Zrozumiesz TV-Sat. Sport*
(Gdańsk: Mariol, 1992)

Bibliography

Baczyński, Jerzy. "Przeciąganie kabla." *Polityka*, supplement: *Polityka – Eksport – Import*, April, 1986.

de Certeau, Michel. *The Practice of Everyday Life*. Translated by Steven Rendall. Berkeley: University of California Press, 1984.

Ciski, Henryk. "Między nami ..." *TV-Sat-Magazyn*, May, 1989.

"co – gdzie – za ile?" *Top*, October 9, 1987.

DJB. "Kluby TV-SAT." *Audio-Video*, no. 1 (1989).

Dolecki, Józef. "Centrum łączności satelitarnej w Psarach." *Audio-Video*, no. 1 (1985).

"Drodzy czytelnicy!" *TV-Sat-Magazyn*, April 1989.

Dulemba, Anna. "Skok na antenę." *Polityka*, supplement: *Polityka – Eksport – Import*, August, 1986.

Fiske, John. *Understanding Popular Culture*. Second Edition. New York: Routledge, 2010.

Fryszkiewicz, Andrzej, Marian Grabski, and Janusz Sarosiek. *Gość czy intruz z kosmosu?* Warszawa: Wydawnictwo Ministerstwa Obrony Narodowej, 1987.

Gorbachev, Mikhail S. *Perestroika: new thinking for our country and the world*. New York: Harper & Row, 1987.

Gorzym, Andrzej. "Jestem dobrej myśli... Rozmowa z Markiem Czajkowskim, właścicielem szwedzkiej firmy PORION AB." *Pasmo*, January 23, 1988.

Grala, Dariusz T. *Reformy gospodarcze w PRL (1982–1989). Próba uratowania socjalizmu*. Warszawa: Wydawnictwo TRIO, 2005.

Henzler, Marek. "U wrót satelitarnego raju." *Polityka*, May 3, 1986.

IBIS. "PIR robi pomiary." *Pasmo*, March 12, 1988.

"Inwazja z kosmosu." *TOP*, December 25, 1987.

Jajko, Krzysztof. "Skrzynki, talerze i transformacja. Młodzieńcze lata telewizji satelitarnej nad Wisłą." *Panoptikum*, no. 15 (2016).

Jamiłowski, A. "Program z satelity." *Pan*, October, 1987.

Kasz, Marek. "Satysfakcja." *Pasmo*, January 30, 1988.

Kobyliński, Seweryn J. "Przygotowanie zestawu odbiorczego." *Ekran*, June 15, 1989.

Kochanowski, Jerzy. *Tylnymi drzwiami. „Czarny rynek" w Polsce 1944–1989*. Warszawa: Wydawnictwo W.A.B, 2015.

Kurek, Tadeusz. *Prawie wszystko o telewizji satelitarnej*. Warszawa: Wydawnictwo Naukowe Techniczne, 1991.

Ładyński, Sylweriusz. "Jak zorganizować osiedlową telewizję satelitarną (3)." *TV-Sat-Magazyn*, July, 1989.

Ładyński, Sylweriusz. "Jak zorganizować osiedlową telewizję satelitarną (5)." *TV-Sat-Magazyn*, September, 1989.

Ładyński, Sylweriusz. "Jak zorganizować własną telewizję osiedlową (6)." *TV-Sat-Magazyn*, October, 1989.

Ładyński, Sylweriusz. "Łódzkie doświadczenia." *TV-Sat-Magazyn*, May, 1989.

MAJ. "Wolno, wolno, wolno." *TV SAT Magazyn*, April, 1989.

Machejek, Jerzy. "Odmówić, odmówić, odmówić...?" *TV-Sat-Magazyn*, March 15, 1989.

Marciniak, A. "Satelitarne vademecum." *TV-Sat-Magazyn*, April, 1989.

Matlak, Krzysztof. "Sława, pieniądze i..." *Głos Szczeciński*, July 18–19, 1987.

Michalewicz, Kazimierz. "Ziemia, kosmos, ziemia. W stacji satelitarnej w Psarach." *Studio*, no. 5 (1975): 32–33.

Misiak, Władysław. *System kulturowy miasta uprzemysłowionego*. Wrocław: Zakład Narodowy im. Ossolińskich – Wydawnictwo, 1982.

Mojkowski, Jacek. "Świat w talerzu." *Polityka*, suplement: *Polityka – Eksport – Import*, March, 1987.

Myjak, Józef. "Bilans w kulturze." *Tygodnik Nadwiślański*, January 6, 1989.

Pawlak, Janusz. "Komu program z satelity." *Nowiny*, September 27, 1988.

Peisert, Arkadiusz. *Spółdzielnie mieszkaniowe: pomiędzy wspólnotą obywatelską a alienacją*. Warszawa: Wydawnictwa Uniwersytetu Warszawskiego, 2009.

Pudlik, Bożena, and Stefan Świętochowski. *Zrozumiesz TV-Sat. Sport.* Gdańsk: Wydawnictwo Mariol, 1992.

SAM. "Telewizja satelitarna: strzał w dziesiątkę." *Pasmo*, February 6, 1988.

SAM. "Telewizja satelitarna w próbach." *Pasmo*, February 27, 1988.

SAM. "Twórca TV-SAT na Ursynowie. Rozmowa z inż. Wacławem Tylawskim." *Pasmo*, June 11, 1988.

SAM. "URSYNAT' zwycięża!" *Pasmo*, May 28, 1988.

SAM. "Wiadomości z frontu TV-SAT." *Pasmo*, June 25, 1988.

SAM. "Wśród serdecznych przyjaciół." *Pasmo*, March 5, 1988.

Sandomierski, Tomasz. "Parada kolosów." *Polityka*, supplement: *Polityka – Eksport – Import*, February, 1985.

"Satelitarna i kablowa TV dla regionu tarnobrzeskiego." *TV-Sat-Magazyn*, August, 1989.

Skwierawski, Franciszek. "Kowalski i komplikacje satelitarne." *Rzeczpospolita*, August 31 –September 1, 1991.

Skwierawski, Franciszek. "Ta bariera językowa…" *Rzeczpospolita*, October 12–13, 1991.

Skwierawski, Franciszek. "Telewizja satelitarna ruszyła…" *Ekran*, June 18, 1987.

Sowińska, Magda. "Pod górkę." *TOP*, October 13, 1989.

Taw. "Wielki kabel na Ursynowie." *Pasmo*, January 9, 1988.

Tł. "Inwazja z kosmosu (1)." *TOP*, December 11, 1987.

Tarkowski, Jacek. *Władza i społeczeństwo w systemie autorytarnym*. Warszawa: Instytut Studiów Politycznych PAN, 1994.

"Telewizja z orbity: Krótki informator TV satelitarnej." *TOP*, September 30, 1988.

Toeplitz, Krzysztof T. "My i nasz system." *Polityka*, June 20, 1987.

"TV-Sat-Magazyn także w CSRS!" *TV-Sat-Magazyn*, June, 1989.

"TV-SAT w Tarnobrzegu." *Ekran*, June 23, 1988.

"TV z orbity." *TOP*, October 7, 1988.

"TV z orbity." *TOP*, December 2, 1988.

"TV z orbity." *TOP*, February 10, 1989.

"TV z orbity." *TOP*, May 19, 1989.

"TV z orbity." *TOP*, June 2, 1989.

Widelski, Konrad T. *Mój telewizor*. Warszawa: Instytut Wydawniczy Związków Zawodowych, 1989.

Zięcina, Radosław. "Mimo braku zasilania studio URSYNAT nadaje." *Pasmo*, May 21, 1988.

Zięcina, Radosław. "Tu studio URSYNAT." *Pasmo*, March 19, 1988.

Interviews

Elżbieta Biwan-Kwiecińska, interview by Krzysztof Jajko, Łódź, October 28, 2015.

Adolf Bogacki, interview by Krzysztof Jajko, Szczecin, September 15, 2015.

Paweł Bruś, interview by Krzysztof Jajko, Łódź, November 27, 2015.

Marek Czajkowski, interview by Krzysztof Jajko, Warszawa, September 16, 2015.

Aleksander Dyl, interview by Krzysztof Jajko, Tarnobrzeg, February 28, 2015.

Jerzy Lubacz, interview by Krzysztof Jajko, Mielec, March 3, 2015.

Andrzej Palma, interview by Krzysztof Jajko, Łódź, October 28, 2015.

Jarosław Piątkowski, interview by Krzysztof Jajko, Tarnobrzeg, March 2, 2015.

INDEX

INITIATING EDITOR
Urszula Dzieciątkowska

REVIEWER
Mirosław Filiciak

TRANSLATORS
Anna Czyżewska-Felczak
Stanisław Krawczyk
Wojciech Szymański

PROOFREADERS
James O'Connor
Kamila Rymajdo
Paweł M. Sobczak

TYPESETTING
AGENT PR

TECHNICAL EDITOR
Anna Sońta

COVER DESIGN
krzysztof de mianiuk

Cover Image: Wacław Pintal

First edition. W.07796.16.0.K

Publisher's sheets 14.1; printing sheets 15.625